BUFFALO
MANAGEMENT
& MARKETING

Jennings is author of
Cattle on a Thousand Hills
Days of Steam and Glory
Greatest Steam Show on Earth
Free Ice Water
Where the Buffalo Roam Again
Farm Steam Shows: USA & Canada
Old Threshers
Slick Trick
Blood on the Killdeer
World's Window
Buffalo History and Husbandry
Old Threshers at Thirty XXX

Hebbring is Executive Director of
the National Buffalo Association
and editor of
BUFFALO! Magazine

BUFFALO
MANAGEMENT
& MARKETING

by Dana C. Jennings & Judi Hebbring

National Buffalo Association
Box 706, Custer, SD 57730
(605) 673-2073

This book includes contributions from many leaders within the buffalo industry

NATIONAL BUFFALO ASSOCIATION
P.O. Box 706
Custer, SD 57730
(605) 673-2073

first printing 1983

Library of Congress Cataloging in Publication Data

Jennings, Dana Close.
  Buffalo, management and marketing.

  Bibliography: p.
  Includes index.
  1. Bison, American. 2. Bison, American—Marketing. I. Hebbring, Judi.
II. National Buffalo Association. III. Title.
SF401.B56J46 1983   636.2′92   83-19516
ISBN 0-918880-03-3   Communication Creativity

Editing, cover and book design, typesetting and printing coordination by:

about
books inc.
saguache, co 81149

To **Fred Matthews**—Texan + Top Hand + Buffalo Herdsman at Custer State Park, South Dakota who pioneered, proved and improved ways to manage this wild, magnificent beast in the wild to prosper humankind.

Thanks a Lot!

So many people helped prepare this book that just to list them all would take a fair chunk of the book.

First, of course, are immediate past President Charlie Tucker and the Board who authorized the project, made the funds available and encouraged it in every way.

Our new president, Armando Flocchini, and his ever-ready sidekick, Jack Errington, get a lot of the credit too. So does founding president, Roy Houck, a major buffalo producer; and Fred Matthews, Custer State Park herdsman. What Roy and Fred don't know about buffalo simply hasn't been found out yet!

Then there are all those other writers and raisers who contributed articles and photos and submitted to interviews. We couldn't have done it withoutcha!

What is the National Buffalo Association?

The National Buffalo Association—actually the membership is international—is a mutual-help society of American Bison producers and others interested in this living symbol of America. Our members' interests span economic, academic and nostalgic areas. Through research, promotion, publications, membership meetings and legislative efforts, the NBA works to prosper the species and the members.

Contents

Preface

Over 900 buffalo books have been written in English that we know of—and we have no idea how many we don't know of. But, to our knowledge, only the NBA's *Buffalo History & Husbandry,* published in 1978, ever told you how to raise buffalo. The present volume gives you more of the husbandry and less of the history.

Again we have consulted not only books but also practicing buffalo raisers. Again we find experiences and recommendations differ . . . doubtless because environments and circumstances, personalities and practices vary. We offer you these differing recommendations that you may choose those that best fit your own operation. Or perhaps you can combine suggested practices for something better. Maybe one rancher's idea will spark a great one of your own.

We suggest you read the book all the way through first, then refer back to specific chapters for specific answers. *Buffalo Management and Marketing* is organized into logical sections. A comprehensive index allows you to quickly consult the pages bearing on your present problem.

We have written for the novice as well as for the expert. This is the one place the beginner can go for bedrock basics; yet this book offers the old hand a lot of good stuff he or she never knew before. We don't intend to answer all your questions about buffalo production. Frankly, they haven't all been asked yet!

Preface

A Birdseye
View

I
I
I
I
I
I
I
I
I
I
I

1
Let's Get Acquainted

These first few chapters are written especially for the newcomer to the buffalo business, giving him or her an overview of the animal and what its culture involves—for "buffalo raising is like raising cattle—only more-so!" according to those qualified in both.

At first we'll just touch on some of the possibilities and pitfalls. Later sections go into more detail on specific subjects from feeding to breeding, management to marketing.

SO YOU WANT TO RAISE BUFFALO

But you're a bit hesitant because you're not all that acquainted with the beast. This is a good time to get into the buffalo business because the whole future of the industry, from ranching to meat to by-products, seems very bullish. There is a strong demand for buffalo meat and by-products (more by far than the industry can supply). Prices are higher than for cattle and going higher. The buffalo industry is still in its pioneering phase, so the potential is immense and untapped.

As in marketing any commodity, much depends on advertising and getting the buyer and seller together. Part of advertising's job is educating the potential buyer to the advantages of your product. Buffalo meat is considered superior to other meats, both in flavor and in health-giving nutritional qualities. The prowess, stamina and longevity of the Plains Indians and early plainsmen were legendary in their own time—and for reasons that modern science is just now beginning to unravel. Food scientists are just beginning to learn that buffalo meat contains less fat and cholesterol—and more protein than beef. Today, fat and cholesterol levels are important to the American consumer, who looks for low levels of both.

WHAT'S IT TASTE LIKE?

Were someone to feed you prime young buffalo meat without 'fessing-up' to what they were about, you'd think it the juiciest, tenderest, tastiest

5

beef you ever ate. "Sweeter than candy and tender as butter" is one apt description. No wild or gamey flavor at all. Buffalo is naturally more tender and tasty than beef, according to taste-tests by a taste-test panel.

Today, there is a large and growing demand for more natural foods. Many people prefer meat that is organically raised—a market still hardly scratched. These people do not want meat from animals fed in high-density pens where they have been stuffed with growth-stimulating chemicals and preventive drugs. Some beef cattle are fed things like sawdust, chicken or cattle dung, and meat scraps from animals condemned for high chemical content or disease. Concerned consumers are able and eager to pay a higher price for a superior product; the pure product that the buffalo rancher can provide.

Do watch out for African and Asian buffalo being sold in this country as "buffalo" meat, or mis-labeled "American Buffalo." The National Buffalo Association (NBA) is working on grading, labeling and policing legislation to avoid such trickery.

NON-ALLERGENIC

Buffalo, unlike other meats, is totally non-allergenic. At least, no report of allergy to buffalo meat has ever come to our attention. Some physicians recommend buffalo meat for patients who cannot eat beef.

MORE PROFITABLE THAN CATTLE FOR THE RANCHER!

The primary financial advantages of raising buffalo over cattle are the higher market prices and lower expenses, due to several of buffalos' superiorities over beef animals. There is less illness, veterinary expense and death loss, due to the buffalos' hardiness. Buffalos practice more efficient feed utilization. And buffalos willingly eat weeds and things that cattle won't consume. Some ranchers report they can stock their grassland more heavily with buffalo than with cattle because of this.

Cattle require a lot of costly grain in the finishing ration, whereas buffalo fatten well on cheap pasture and hay. Some producers give buffalo a light finish-feed of grain for topmost quality, but do not spend nearly as much as is necessary for finishing cattle.

Buffalo on the hoof sell for 20–50% more than beef animals. The demand for live animals, meat and by-products exceeds the supply and is expected to outpace production for years to come. The buffalo rancher who studies the market and advertises his wares should have no difficulty getting good prices.

THE NAME OF THE GAME IS MEAT

The end product, of course, is meat. It is not expected that buffalo will replace beef for many generations yet. The total buffalo population of the U.S. today is estimated at 75,000 head, compared to a cattle/calf count of 120 million or so. About 10,000 buffalo go to slaughter each year. Compare this with the 133,000 cattle that are slaughtered every DAY! About the only buffalo going to slaughter are the surplus bulls and non-productive cows. This is because raisers are still building their breeding herds. Therefore you can get a higher price for your breeding stock than for slaughter animals. Prime customers are fine restaurants, clubs, the gourmet trade and people whose physicians have put them on a buffalo meat diet. Meat markets and stores seldom offer buffalo simply because they cannot get a steady supply—there just isn't enough. When they do get some, they make a big feature of it to garner all the publicity (and profit!) possible from this unique romance item.

Thus you see, there is little fear of buffalo meat flooding the market. With increasing consumer acceptance, the demand for breeding stock should continue strong for many years, and the meat likewise—each fire fueling the other.

BY-PRODUCTS ARE MONEY MAKERS!

Don't overlook profits from the bits and pieces. As this is written, green hides are fetching $50 to $100, mounted heads up to $2,000. Bleached skulls bring $75–$100, horns $5 to $15 each, patches of buffalo leather $4 a square foot. Even the tails, hooves and toenails have their price!

MORE GROSS—LESS COST=MORE NET!

Nobody in a state of sobriety ever advocated buffalo raising as a royal road to riches. It's an enterprise for the long pull, for the rancher of true grit, willing to swallow some dust and take some lumps—not for the in-and-outer. But for the enterpriser who's in it for growth, buffalo are a better deal than cattle.

There is a higher gross. Buffalo consistently bring 20 to 50% more than beef—on the hoof or on the hook. The buffalo market is less subject to the boom/bust gyrations of cattle.

A foundation herd may cost you more than a cattle herd, or maybe not, depending on the market that day and how good a horse trader you are. Fencing costs more. Corral systems and handling equipment cost much more. After that, buffalo cost less to raise because they're more efficient foragers and feed converters and they require less grain for finishing. They're also hardier, socking you with less veterinary expense and death loss.

As one rancher told us, "Buffalo paid for our cattle these last few years!" The buffalo market, both breeding stock and meat, maintains a steady upward trend, without the cycling of the cattle market. So while this is no "get rich quick" scheme, buffalo raising has a bright and stable future.

2
Raising Buffalo
for Fun and Bucks

Buffalo can be an expensive hobby. People who can afford it like to keep a few head around to give the place a Western flavor, or as self-propelled conversation pieces.

Operators of tourist attractions learned long ago that nothing pulls traffic off the highway like a pair of buffalo grazing nearby. They also learned that buffalo are not loners. One animal by him or herself does not do well. Operators typically keep at least two. And—given the proper gender—you know what that leads to!

Most importantly, buffalo can be a major money-maker for the rancher who has the grit to stay with 'em. This is not a game for the in-and-outer. The profit potential is definitely there, but it must be nurtured.

RANCHING PROFITS

Nobody ever got rich off of one or two head. The real money to be made is in ranching—whether by hands-on, boots-and-saddle ranching or by checkbook ranching.

The newcomer to the industry should give serious consideration to purchasing calves at the outset. Calves adapt better to a new home and will be far less likely to roam, as mature buffalo sometimes do. You'll have to wait a couple of years before you're actually in the business when you start with calves, but the potential grief you save yourself will be worth it! Adult animals, when introduced to a new location, have a tendency to leave home, since they are confused and don't know where home is. Also, female adult buffalo often take a year or two to settle down and start producing calves again after they have been moved. Since you're probably going to have to wait for calves anyway, you might as well start with calves and let them grow up on the place.

By purchasing young stock, you are also far less likely to buy someone else's problem. A troublemaker or a wanderer is going to be a problem no matter where it lives, and it's going to teach your other buffalo its bad habits. About the only solution is to eat it. This is one of the nice things about the buffalo business: you can eat your mistakes!

After your herd is established, then it's okay to introduce new adults. As long as you haven't bought a bunch of rabble rousers, you shouldn't have any problems. The newcomers will stick with the oldtimers. If you have land, fences, horses, capital and all that good stuff, you have numerous money-making opportunities in buffalo. *If* you take the time. But just because you don't own a whole big bunch of land or a whole big bunch of money, despair not. There are other ways, as we will shortly tell you.

TAX ANGLES

Taxes can make a bigger difference than weather and markets, and are just as variable, and even more unpredictable! It has been said that all you need to get rich in agriculture these days is a big spread and a good tax accountant.

To give here a detailed recommendation on handling your buffalo taxes would be an exercise in futility, because next year the tax laws will change again. Watch *BUFFALO!* magazine for continuous updates. Meanwhile, make friends with a competent tax consultant.

The first thing to do is to sit down with your tax accountant and figure the angles. Foundation-stock purchase makes you eligible for depreciation allowances and tax deductions that can save you more taxes than the purchase price. Ponder fast tax write-offs.

Deduct most of your expenses as current while your herd is growing to commercial scale. Then when you peddle breeding stock, even though the animals have been fully depreciated, claim capital gains.

Buffalo need be owned for only one year and can be sold and the proceeds can come under capital gain. Contrast this with the two-year requirement for cattle. When the law was changed, NBA persuaded the lawgivers not to change the status of buffalo. Also, you can take a 7% tax credit on buffalo purchases. The depreciation schedule runs from 10 to 12 years. While you're investing in buffalo, the IRS can't touch those funds.

HOW TO BE HAPPY (THOUGH A TAXPAYER)

There are generally two methods of accounting for a farm or ranch, which would include a buffalo breeding herd.

On the CASH BASIS, all of the expenses, except capital expenditures for equipment and purchased buffalo, would be deductible. Income would be reported only as received. Depreciation can be taken on any purchased buffalo, usually over a seven-year period. Sale of the purchased buffalo would be subject to depreciation recapture rules but otherwise would be a Capital Gain.

On the ACCRUAL BASIS, the buffalo rancher must inventory the direct costs of raising the animals, either home born or purchased. Depreciation is then taken on a seven-year period when the raised animal reaches the breeding age. Any sale of the breeding herd, if held 12 months, would be a Capital Gain, except for the recapture of depreciation. It is normally considered better not to inventory the raised animals but rather to treat them on a cash basis.

Individuals and Sub-Chapter S Corporations must take care that the IRS does not treat any losses as Not-For-Profit-Activity Losses or "Hobby Losses." These losses under certain circumstances could be disallowed if there is not any profit shown in two out of five or more consecutive years.

IMPROVING ON NATURE

One way to sneak ahead of nature is not to wait for your calves to grow up and start raising little ones of their own. Rather, trade them (tax-free) for mature cows. You can start with 10 cows and one bull. By trading the resulting heifers—two of them for one cow—you immediately boost the young-bearing potential of your herd, as the cows can become pregnant the next time they come into heat.

Ways To Go

1) RANCHING: Buy a ranch and a foundation herd and start raising buffalo. (It's a good idea to hire an experienced buffalo herdsman at the outset—and they're hard to find.) You'll be lucky to get 1,000 acres and 100 head (the minimum economic unit, most places) for less than half a million. And you can expect three to eight years to flit by before you get any return. Even old hands with cattle have a lot to learn about buffalo. Ranching offers the greatest long-term potential; it also carries the greatest risk. NBA can help you locate land, herds and experienced managers/operators.

2) MINI-RANCHING: If you've got a few wasted acres, buy one bull and 10 heifers or cows (to be on the safe side, provide two bulls for 10–15 females). A heifer herd and yearling bull entail a wait of one to two years for calves. Meanwhile, you're deducting.

3) CHECKBOOK RANCHING: If real bumps-and-lumps ranching is not for you, you can be a checkbook rancher. To do this place your animals with a working rancher at a contracted fee-per-head-per-year, or for a share of the increase. Or buy so many animals out of a working herd and leave them there. Another way is to buy shares in a ranch corporation.

Sale of breeding stock, as we said, comes under capital gains, even after depreciation. But sale for meat, trophies, etc., is fully taxed.

4) STOCK SHARE OF PARTNERSHIP. NBA can help you find a rancher looking for fresh capital. Here you get the benefit of a going ranch plus the skill of an experienced rancher. One option might be to buy shares of stock in his corporation. Another is to form a partnership with him.

5) *"PUTTING OUT" BUFFALO:* You buy some critters and a working rancher raises them for you on shares of the progeny or for a cash fee. The agreement can include his locating, buying and hauling the animals, after you lay your money down.

6) *CUSTOM FINISHING:* Buy young bulls and put them into drylot for finish-feeding. Make sure the operator has a successful record with *buffalo,* not just with cattle. Buffalo are a whole new ball game, with different rules and a new pitcher. Your neighborhood feeder will fatten them for you on a flat rate, per-pound or percentage contract. Or, simply buy a share in buffalo already on feed. Another alternative is to provide a feedlot operator with a nice fresh bunch of capital for a slice of the take.

The NBA can help you locate the components of any of these plans.

UNEASY CREDIT

Banks and other financial institutions have demonstrated an uneasiness about lending money for the purchase of buffalo. The nature of the animal, and stories about their enormous and potentially damaging strength, affect banker's judgment as to their suitability as collateral or as an investment. Because the market for buffalo is much more restricted and extremely dependent on the law of supply and demand, the risk aspect fluctuates widely from one area to another depending upon the marketing ability of the producer. Without a robust, ready market on a consistent basis, Durham Meat Company says that those raising buffalo can anticipate limited financing without other collateral and personal guarantees.

MARKETING

Here comes the bottom line. Do they pay? How much?

There have never been enough buffalo to establish a regular market. This is a small disadvantage, however, because most growers can sell more breeding stock and more meat at fancy prices than they can produce anyway. This is one commodity the producer has some say about the price of. Buffalo aren't like cattle and corn, where you haul the stuff to market and take whatever somebody wants to give you. When you're selling buffalo, on the hoof or on the hook, the buyer comes to you and says, "How much do you want?" and *he* pays the freight!

NBA is planning a computerized clearinghouse setup to bring buyer and seller together most expeditiously.

Some people worry about saturating the market. Considering that figures from the USDA Economic Research Service show that the U.S. slaughters some 133,000 head of cattle (not counting calves and on-farm) *a day,* and only about 10,000 buffalo *a year,* buffalo are a long way from saturating anything.

NO LONGER ENDANGERED

Some people think buffalo are extinct and react with real surprise to learn some still live. Others think they still teeter on the brink of extinction and it's an ecological crime to eat one. Part of your job as a buffalo rancher is to convince people of the following:

Buffalo are not an endangered species, thanks to far-sighted pioneering growers. The government got into the act only after the ranchers had won the battle. There are now over 75,000 head in the U.S. alone, more in Canada.

Approximately 50% of the calves born are bulls, and we need only about one out of every 10 of those. If we don't eat those surplus bulls, what can we do with them?

By buying buffalo meat you help support the continuing effort to re-establish the species and make them more available.

FUTURE MAGNIFICENT!

As population increase and resource scarcity continue on their collision course, the buffalo's more efficient feed conversion and ability to forage where cattle can't go will make him even more important as a national resource. And if buffalo make the contribution to human health current research promises, the buffalo will become as important to our culture as he was to the Plains Indians.

BUFFALO CAN BE PROFITABLE

These figures are estimates. There must be variations due to location, climate, operational costs, land values, carrying capacities, etc. And inflation is a long-term imponderable. This uses an original unit of 10 heifers and one bull calf; figures can be used in multiple. Theoretically, at least, larger numbers introduce economies of scale. Land values vary so widely, you'll have to insert your own land numbers.

Investment in corrals, fences & equipment.................$10,000
10 heifers @ $500....................................... 5,000
1 bull calf 500
Cost of running these calves 2 years @ $100/hd/yr......... 2,200
Taxes, vets, misc. 2,000
MAKING A TOTAL INVESTMENT OF \$19,700
The 3d year, with an 80% calf crop you should have 8 calves
    valued @ $500 @ 6 months, showing a crop return of.......4,000
$4,000 income on a $20,000 investment is 20% return after 3 years
OR BETTER YET, KEEP YOUR BULL CALVES TILL THEY'RE
    2 AND SELL THE MEAT
Sale of heifer calves (based on 50/50 sex ratio) in 3d, 4th, 5th years
    12 heifers @ $500 6,000

Value of meat from 4 2-year-old bulls @ $1,000 4,000
Value of heads, hides & misc by-products 1,000
Upkeep & misc. in years 3, 4, 5 6,400
TOTAL INVESTMENT & EXPENSES AT END OF
　　YEAR 5 ....$26,100
Total return at end of year 5........................... 11,000
Your income at the end of year 5 has now totaled 40% on your original
　　investment, including operating expenses, AND REMEMBER you still
　　have 4 yearling bulls and 4 bull calves—this is money just waiting to
　　be deposited in the bank in your 6th & 7th years.
OR IF YOU DECIDE TO RAISE THOSE HEIFER CALVES TO
　　ADD TO YOUR BREEDING HERD, IN 10 YEARS THIS IS
　　WHAT YOU'LL HAVE:
50 cows in production and you'll have sold about 25 excess 2-year-old
　　bulls for meat with a return of about $31,250
You'll have replaced your older bulls with younger bulls and sold the
　　big trophies to hunters for about $2,500 each (×4)........ 10,000
By year 10 you will have long ago recovered your original investment.
You will have 50 cows producing 20 heifer calves annually
　　@ $500 10,000
You will have 50 cows producing 20 bulls annually @ $1,250 (meat &
　　by-products)....................................... 25,000
You will sell your large trophy bulls @ 2 or 3 per year
　　2 trophy sales per year will realize a return of 5,000
Herd maintenance 5,000
Other buffalo related expenses should not exceed........... 4,000
NEW INCOME $31,000
Maintenance and upgrading of your handling facilities, disease control
and unforeseen disasters can take a big bite out of the projected net
income, but then again your production should be higher than our very
conservative estimate of 80% (most get 90–95% under ranch conditions).
And you can look forward to ever increasing prices for meat, by-products
and breeding stock. Remember—we never said buffalo fit a fast buck
scheme.

3
Your
Questions
Answered

The American buffalo is not a buffalo.

It's a bison.

The true buffalo is the Asiatic water buffalo, *Bubalus bubalis.* The African or cape buffalo is zoologically designated *Sycerus caffer.* The American so called buffalo is truly *Bison bison bison,* or bison for short. It's flesh is a far superior viand to either of the others. As used in this book, "buffalo" refers to *B. b. bison* unless otherwise noted.

B. b. bison is native to North America, having descended from a larger animal that apparently originated in Asia and crossed over thisaway while the early horses and camels went thataway. It was hunted to extinction by early Indians.

BISON HYBRIDIZATION CHART

Buffalo and cattle are, you might say, cousins. They belong to the same family and tribe. The only closer living relative is the European forest bison, the wisent (say VEzunt).

Taxonomic Classification	American "Buffalo"	Wisent	Cattle
Order	Artiodactyla	Artiodactyla	Artodactyla
Suborder	Ruminata	Ruminata	Ruminata
Family	Bovidae	Bovidae	Bovidae
Tribe	Bovini	Bovini	Bovini
Genus	Bison	Bison	Bos
Species	bison	bison	taurus
Subspecies	bison	bonasus	

B. bison and *Bos* hybridize only with difficulty, and the males are sterile.

B. b. bison and *B. b. bonasus* cross easily.

15

Cattle have 13 pairs of ribs; *B. b. bison* has 14—an extra pair on which to hang high-priced cuts! The wood bison of Canada is thought to be the same subspecies as the "mountain bison" reported by frontiersmen, and is designated *B. b. athabascae*. The latter is nearly extinct as a species since it was unwisely mingled with the prairie bison in the 1920's and they promptly interbred. Today it's hard to distinguish a Canadian bison as being *B.b.b., B.b.a.* or a mixture, outside of a recently-discovered isolated pocket of *B.b.a.*

BUFFALO NATURE

Buffalo are a different natured animal than cattle. People who have worked cattle for years, and can predict what they're going to do, find they can never outguess a buffalo. In working buffalo, you soon develop the philosophy of leading, not driving. Remember the buffalo axiom: A bunch of cowboys can herd buffalo anywhere the buffalo want to go! These critters have a stronger herd instinct than cattle and stay in more cohesive groups.

Give your buffalo plenty of good feed and water, salt and minerals, and they'll seldom get breachy. Once when floods took out a lot of Roy Houck's fences on the Standing Butte Ranch (Roy was NBA founding president and is patriarch of the family that owns and operates one of the biggest buffalo ranches in the world), his cattle scattered. "The buffalo stayed put. They know when they've got a good thing," comments Houck. Buffalo have been observed staying put within rickety fences that wouldn't hold a sick milk cow. As a rule, buffalo are not as hard on fences as cattle are. But don't count on it. Some owners have a hard time keeping their buffalo confined even with good tight, strong fences.

Beware, there's no such thing as a tame buffalo! We've heard of buffalo bull calves raised as pets, which at about age six, killed their masters.

Buffalo are wild animals, no matter how tame they may seem. There have been no new animal species domesticated for thousands of years. We guess that's because only the wild and untamable ones are left. If you want to make a pet of a buffalo bull, better cut him or turn him loose at about five years old.

A point that can't be stressed too often is that buffalo are dangerous! People tend to become complacent with anything that grows familiar. But for everyone's safety, every buffalo raiser should take precautions. Buffalo are unpredictable. Even when raised as pets, they still remain wild animals.

The ideal location for buffalo is back away from people. The less the beasts are handled and annoyed, the better for animals, handlers and spectators!

Ranchers who are into buffalo commercially know this. On the other hand, many of the people who keep a couple as pets or curiosities are

Bill Moore, Bar X, shows cattle fence adapted to buffalo by post extenders and offset hot wire.

Used materials save money on fencing. Bill got a truckload of chain-link fencing cheap at a distress sale.

unaware of the dangers of keeping buffalo in built-up neighborhoods, in small enclosures, behind flimsy fences. A couple of buffalo on an acre just outside of town can be a fun thing for the owner and the townsfolk alike. But the owner should be aware that the only thing keeping the buffalo in is faith. If buffalo take a mind to it, there are few fences that can hold them. A loose buffalo in suburbia can be a real hazard. The owner better be well insured, or prepared to change his name and leave town without a forwarding address.

We have an Oregon report of a buffalo that wandered 250 miles . . . a record, so far as we know. This animal could have done a lot of damage. Fortunately for the owner, it did very little, although a landowner got mad and shot him. The animal's owner did have a lawsuit out of this escapade.

The buffalo hobbyist should keep these incidents in mind. Our purpose is not to discourage potential buffalo raisers or even pet owners, but to advise them of the risks involved. Chances are, you'll have no problems, but be aware of the possibilities and have a plan of action ready just in case. In the meantime, enjoy your buffalo. They are magnificent animals!

Buffalo ranching is like cattle ranching, only more so. They are bigger and stronger than cattle and more likely, when pushed, to take a flying leap for freedom, or to imitate a bulldozer. The buffalo rancher needs stronger, higher fences, corrals, chutes and gates. Also trucks.

WHAT DO BUFFALO EAT?

Grass. Weeds. Brush. Hay. Grain. Generally, the same feed that cattle eat, except adult buffalo often refuse grain if they've not been accustomed to it as calves. They are less picky and will eat some forbs that cattle won't.

HOW MANY ACRES PER ANIMAL UNIT?

Generally, the same as cattle in your area—which may range from one acre per head in the humid East to 100 acres per head in the arid West.

In general, any pasture that's good for cattle is good for buffalo. They eat some weeds and brush that cattle won't, but do skip loco weed. For this reason, you can sometimes pasture more buffalo than cattle on a given area. Veterinarian Ken Throlson discusses this in a later chapter.

HEALTH

Under natural range conditions, buffalo are much more resistant than cattle to most diseases and parasites. Crowded into tiny pastures or filthy pens, where they are forced to recycle their excretion, they do get sick.

When maintained under close to natural conditions (lots of room to roam), buffalo constantly move as they graze, traveling miles each day. Thus, the parasites, eggs and larvae in their droppings will have been killed by sun and rain before the animals re-graze a given area. However, most of you will have to maintain your buffalo on limited range, compelling them to re-graze a given area before nature has had time to do her purifying work. Under these conditions, they will pick up the little monsters still lurking in the grass, making parasite control a must. Later we'll tell you how to cope with that too!

Fly control is a good investment, since an animal fighting and feeding flies is not gaining weight. Some ranchers vaccinate against local diseases.

HOW LONG DO THEY LIVE?

Normally 20 to 30 years, which is two to three times the lifespan of cattle. Buffalo cows have been known to calve past age 40 ... but don't plan on it. Most ranchers sell off both bulls and cows after eight or nine years to help insure a good calf crop.

BREEDING

Buffalo reach puberty a year later than cattle, hence drop their first calf a year later, usually at age three. Although nature supplies a bull for every heifer calf, a working ratio of 1:10 is best in a big herd, and 2:10 to 2:15 in a small herd.

Gestation is about the same as for cattle: 285 days—a little over nine months. Late summer is the rutting season. Most calves drop in April or May. A herd is much easier to manage if calves arrive in a bunch.

CROSSBREEDING

Don't.

The idea of combining the buffalo's hardihood with the cattle's docility has fascinated breeders since buffalo were first discovered circa 1500. Many attempts have been made with extremely high maternal/infant mortality. First cross males are sterile. The consensus is it's not worth the trouble and expense.

Don't try it unless you're able and willing to absorb a huge loss of mothers and babies. Small cattlemen have been wiped out trying to use buffalo bulls or semen.

WHAT IS THE PROFIT POTENTIAL IN RAISING BUFFALO?

Immense.

Your foundation herd shouldn't cost you any more than a good cattle herd. You'll need higher, stronger fences and corrals, so this investment will be a third to half again more than for cattle. After that, your costs for feed, veterinary, etc., will be lower, with less death loss. Housing costs are zilch, because even if you build them a nice shed, they won't go in it, preferring the natural shelter of rough terrain, thickets and trees.

Buffalo offer huge long-term gains to the rancher, the feeder and the investor. Demand for buffalo meat and by-products continues to out-strip the ever mounting supply.

SUPERMOTHER

The domestic cow (*Bos taurus*) has been called "foster mother of the human race" because she converts cellulosic materials such as grass, weeds, etc., into milk and meat for us. It is her four-chamber stomach and cud chewing that make her so. Bison, via similar stomach and cud, can convert even rougher roughage into good things for people.

Cud chewing is common to cattle, buffalo, sheep, goats and camels. It is cud chewing, or rumination, that gives this class of livestock their generic name, ruminant. The ruminant doesn't chew it's food at first—it simply bites it off and swallows it. Later, lying in the shade and thinking deep thoughts, it regurgitates wads of feed and chews them, called ruminating. Here's the process, step-by-step, without which the human race could not survive . . . at least not in its present numbers.

1) The fresh bitten forage lands in the first chamber, the rumen or pouch, which may hold as much as a gas barrel. The billions of micro-organisms in this moist warmth feed on this material, breaking down the tough celluloses and lignins and producing the gas which must be got rid of by belching. If the belch reflex fails, bloat results. This is a serious condition in cattle, but the word is that buffalo do not bloat. As usual, when people say "buffalo never—" or "buffalo always—", we hedge and say, "Well, under normal conditions—."

2) The saliva-soaked feed passes into the second chamber, the reticulum or honeycomb bag of mucous membrane folds forming hexagonal cells like a honeycomb, where more fermentation proceeds. Contractions of this organ further mix the food with saliva and water, forming balls about the size of your fist.

3) Watch Bison or Bos at rest. You'll see a spasm or hiccup flick the flanks and diaphragm. You'll see a bolus slide up the gullet into the mouth. Here it gets a second chewing, this time thorough, before being swallowed again. This time it drops into the third compartment, the omasum (also called the manyplies or psalterium from it's many folds, like the pages of a prayerbook), or it may be recycled for a second rumination. (Note: Grain is not ruminated, but passes directly to the third compartment on initial swallowing.)

The omasum's function is not well understood. It is known that it divides the cud into smaller particles and extracts excess water from the food and re-absorbs it.

4) Then on to the abomasum (also called the reed or rennet). This is the true stomach, the only compartment that secretes digestive juices, the only one where true chemical digestion occurs. It is next in size to the paunch. (The lining of this stomach, or the crystallized enzyme rennin

20

made from it, is the cheesemaker's material for curdling milk, although today other enzymes and acids are also used.)

In the calf, the abomasum is the largest compartment and digests the milk. Only when the calf begins to graze do the other compartments begin to develop. The above discussion is based on domestic ruminants. We assume that buffalo rumination works along the same lines.

THERE IS LOTS WE DON'T KNOW

Hardly any basic research has been done on the buffalo. The *Biology Data Book*, for example, the biology business' bible, gives you blood counts and intestine lengths and brain weights for man, cow, cat, mouse, camel and bat, but doesn't list bison. Practically no basic biological study has been done on bison—one reason being that they are decidedly more expensive and rambunctious than mice!

4
Will
They Adapt
To My Area?

The Plains bison's original range stretched east of the Rocky Mountains from the Northwest Territories into Mexico. They ranged as far east as New York to the north and Florida to the southeast, right to the Atlantic coast at some points. Wood bison inhabited the continent west of the Rockies, but apparently did not reach the Pacific, according to Tom McHugh in *The Time of the Buffalo*.

This continent's early settlers found buffalo nearly everywhere (although there is no record of the colonists of New England ever encountering them). This explains the many towns, counties, rivers, creeks, gaps and hills in the East named for buffalo. Buffalo, NY, arose on a major migration trail that can be tracked through Ohio, Kentucky and down into Georgia. Buffalo numbers in the East never approached the vast herds on the Great Plains, but there were enough to cut migration trails six feet deep and many feet wide in some areas.

As the pioneers pushed westward, the buffalo fell before them. By 1700, few buffalo could be found in the East and by 1750 they were all gone. The huge herds west of the Mississippi remained, little disturbed, for another 100 years. Pioneer accounts describe them as extending "from sky to sky," and "in numbers numberless," and "the whole country was as one robe." Thundering herds took days to pass. People who saw them estimated the continent's buffalo population at 120 million (about equal to today's U.S. count of cattle and calves). Easterners who hadn't seen them allowed only half that many.

Prior to the railroads, there was no "southern herd" and "northern herd." There was just one huge herd stretching from central Canada into Mexico. People of the time believed all buffalo migrated annually from Canada to Mexico and back. Later studies suggest that any given buffalo would travel three or four hundred miles in a sort of circle in a season.

Numbers were so vast that no one, except a few Indian sages, could

23

conceive that they would ever be gone. The commercial hide hunters sallied forth with their well equipped wagon trains in the spring of 1883 bent upon the annual harvest of hides, as they had been doing for the past two decades. But they returned empty and broke. The impossible, the inconceivable, had happened: the buffalo were gone!

Historians generally agree that the buffalo population hit its low point in 1900. Nobody knows how many were left in the world, but it is fairly well agreed there were less than a thousand in the few ranchers' herds, ones and twos owned by exotic animal fanciers, a few in zoos, a Pablo-Allard herd in Montana, and a little herd of Wood bison in Canada. Poachers left only six alive in Yellowstone.

Some of the earliest herds to be re-established in traditional buffalo country got their seed stock from non-traditional areas. The Wichita Mountains Wildlife Refuge near Cache, OK, got its start from buffalo shipped from the New York Zoo in 1907 by the New York Zoological Society.

Buffalo came to roam Santa Catalina Island off the California coast by accident. Buffalo were moved onto the island in the 1920's for filming a movie based on a Zane Grey book. After the film was shot, attempts to round them up failed. The small herd throve. About 10 years later, new blood was introduced from the Sherwin ranch in Colorado, which herd had been founded from the Scotty Philip herd in South Dakota. They in turn came from a few calves salvaged by Fred Dupree on a freight run into Montana in the 1880's. Today the Santa Catalina Island herd is maintained at about 450 head.

Since the 1960's, through private buffalo enterprise, these critters have been re-introduced to their original range and introduced into many places they never dreamed of. Human ingenuity has placed them in many non-traditional areas. Being hardy, adaptable animals, buffalo today thrive in many locales unsuitable to domestic livestock. Herds thrive from Michigan's cold, wet Upper Peninsula to the subtropics of southern Florida; from the west coast to the east coast; from Alaska's Kodiak Island to California's Catalina Island; from Alaska's tundra to Colombia's jungles; from Hawaii to West Germany to Japan. It seems buffalo will adapt to nearly any climate or forage. The only reason they're not in some places is because nobody's opened the tailgate and let them out there!

Buffalos' adaptability to harsh cold is historic fact. Between 1906 and 1908, 75 cowboys herded 700–900 head of Plains bison to northern Alberta where an area expected to maintain up to 10,000 head was reserved for them. They multiplied until there were so many that the Canadian government decided to ship the herd to the Northwest Territories in the Great Slave Lake country where the resident herd of Wood bison wasn't faring so well. The Plains bison settled right in, bred indiscriminately with the native Wood bison and by the time there were 10,000 of them, it appeared the Wood subspecies in the area had been over-whelmed and absorbed by the dominant Plains species. The Northwest Territories herd has fluctuated from 15,000 to 1,000 and has settled down to a stable 5,000, more or less.

In northern Ontario, a small herd of Plains bison was introduced from Alberta in the early 1930's. By the 1950's, this herd numbered about 100. It thrived in an area where the only browse was blueberry bushes and other shrubs, until poaching wiped it out.

More recently, buffalo were planted on Kodiak Island, off Alaska. This is one of Alaska's milder climates, where the grasses offer poor nutrition and Kodiak bears are a hazard. But the buffalo are holding their own against both the poor forage and the hungry bears. (Hunters consider the Kodiak bear to be the fiercest animal in North America, making the grizzly look like a deflated cream puff.)

No less remarkable, but far less documented, is the buffalo's adaptability to extremely hot climates. Buffalo are thriving in parts of Florida where only the tropical Brahma does well. In the 1980 summer, when the plains from southern Kansas to Texas suffered weeks of 100° + heat, cattle died by the herd and buffalo survived. The ordeal took its toll in the animals' condition, but as soon as the hot spell passed, they snapped back.

The buffalo's ability to acclimate to extreme heat has been attributed to it's ability, unlike cattle, to sweat. But that doesn't really explain this phenomenon. Buffalo, though they have "open pores," don't really sweat as horses and people do. They get wet under highly stressful situations, such as handling, but their "sweat" is more oil than water and occurs even in sub-zero weather.

When you run a buffalo in hot weather, his tongue will hang out until it almost brushes the ground, but I doubt you will see him sweat. It seems to be excitement, fear, and nervous tension that bring out this exudate, whatever it is.

What traits make buffalo adaptable to hot climates are still unknown, but they seem to do as well in the tropics as in the far North.

Buffalo not only acclimate to differing climates, they also evolve quickly to the outside influences brought to bear on them. Southern buffalo, such as those in the Wichita Mountains and on Catalina Island, where they've had 20 or more generations to adapt, change in form. They become smaller, narrower. A large adult bull on Catalina Island will tip the scales at 1,400 to 1,500 pounds, compared to over a ton for his northern brother. This difference is theoretically explained that buffalo in warm climates don't need the huge body mass or broad back required in cold climates to absorb and store heat from the winter sun. It is a mathematical fact that the larger the body, the less surface per unit volume to lose or absorb heat.

This same ability to adapt is some scientists' explanation for the development of the Wood bison variant. Genetically, Wood bison are identical to Plains bison. Only outward characteristics distinguish the two at this stage in history. When more genetic data is in, perhaps a genetic difference will be found, but for now, Wood and Plains bison cannot be distinguished by blood typing.

Outwardly, Wood bison are taller, with less pelage. They lack the long beard and pantaloons, as that hair is a handicap in the bush. This evolution of traits due to outside influences apparently carried through for enough generations that the traits constituted an identifiable subspecies. These traits hold true even after generations of living in the Plains environment.

The Plains bison, introduced to the Wood bison range in the Northwest Territories, have shown no indication of turning into Wood bison, even after many generations. The explanation still eludes us because geneticists have proved that acquired characteristics are not inherited.

5
Getting
Started

First, decide if you're going to be a full-fledged rancher, a mini-rancher, a hobbyist or a checkbook rancher. Your decision will depend on how much land, capital, time, experience, and determination you have. If you think buffalo might be nice to try for awhile, throw this book away right here (sorry, no refunds!).

More than one millionaire has bought a herd, trucked them to their new home, and discovered problems beyond himself, beyond all his horses, and beyond all his men. So there he was, with a rebellious and rapidly deteriorating herd tearing up his nice new corrals. And it takes time to find a buyer. Distress sales are no fun, especially when both the buffalo and their new owner are distressed.

STEPS TO FOLLOW

To the buffalo beginner we suggest these steps, in this order:

1) Read this book from cover to cover.

2) Between chapters, contact the National Buffalo Association, P. O. Box 706, Custer, SD 57730, phone 605-673-2073. NBA is the one international source of buffalo information.

3) While you have them on the phone, take out a membership or at least subscribe to *BUFFALO!*, their international bimonthly magazine. It keeps you up to date on the buffalo world, advises you of animals for sale, sources of equipment and supplies, plus shares other buffalo raisers' experiences and expertise. NBA is the only clearinghouse that brings buyers and sellers together. A buffalo herd book (registry system, similar to those used for purebred cattle and other domestic livestock) is currently being developed.

HERE'S EVERYTHING KNOWN ABOUT BUFFALO GENETICS

If there's any scientific information on buffalo genetics, our research failed to turn it up. They're all pretty much alike, so presumably are homozygous—genetically similar. Most wild animals are that way: see one cottontail and you've seen 'em all. This is expected in buffalo because all trace their ancestry to the thousand or so individuals left in the world in 1900.

Crossbreeding with cattle is so difficult that it is doubtful there is any hidden dilution with *Bos* blood. Individual differences are thought to be due more to environment and nutrition than to heredity. The best advice so far is . . . "breed the best to the best and eat the rest."

We know of one herd—that on The Little Buffalo Ranch 40 miles southwest of Gillette, WY, that is totally inbred. Every animal is descended from one bull and either of two cows that founded the herd in 1925—no new blood has been added. And it's one of the nicest looking herds around. This proves what geneticists have long known: inbreeding concentrates the traits, both good and bad. If you've got all good traits to start with, and peel the bulls off the bottom (that's rancher talk for culling out all but the very best bulls), inbreeding is a safe way to go. It eliminates the chance of introducing disease or undesirable traits. As Dr. Throlson says in a later chapter giving advice on buying, you can't make champs from cheapies.

4) Decide right now on type. Do you want to breed big, long, rangy buffalo—or short, compact ones with a nice big hump? Do you want dark hair or light? Then buy that kind for your foundation stock.

BUYING YOUR BULL

5) Bulls, of course, are no problem. Nature provides 10 for each one needed. There are always plenty of bulls for sale. When you go shopping try not to buy the other feller's problem. Look out for the bull he wants to get rid of because it's mean—a fence wrecker—impotent—diseased, etc. Ask some probing questions.

Pick the bull that's the biggest for his age, provided he's got that macho look about him: the masculine head, the Boone & Crockett horns, the bulldozer tread. Aim for the one at the top of the pecking order, the one all the other bulls step aside for. Pick the greediest—the one that eats the most and horns the others away from the trough. Choose one a cattleman would call "typey." Make sure he's sound—no gimps or sores, no droopy ears, no sore teeth. Look for a bright eye.

COWS

6) Many a fledgling cattleman got his start buying some decrepit old grand dams with maybe one or two calves left in them—because they were cheap but carried champion genes. That's something to consider if your wallet is Twiggy-thin.

On the other hand, grannies are sot in their ways and may not take a liking to your new fangled notions. Young'ns come higher, but have more good years left in them, although until they have brought forth

BISON JUDGING CRITERIA

Score Card	Max	Poor	Good	Excellent	Superior
Character	40	0–25	26–30	31–35	36–40
Soundness	10	0–4	5–6	7–8	9–10
Condition	10	0–4	5–6	7–8	9–10
Size	20	0–10	11–15	16–18	19–20
Quality	20	0–10	11–15	16–18	19–20
TOTAL	100	0–53	58–72	77–87	92–100

Characteristics	Bull	Cow
Weight	+2000 pounds	+1000 pounds
Height at shoulders	5.5–6.5 feet	4.5–5.5 feet
Hump	Massive, sometimes very angular	Smaller, more rounded
Head	Broad, triangular appearance	Narrow, linear frontal appearance
Horns	Proportionally thick at base, simple curve	Proportionally thin, often recurved in side view
Head hair	Thick, long "mop" on top, lengthy	Less dense, shorter
Beard	Lengthy	Relatively short
Cape & chaps on forelegs	Lengthy, pronounced	Relatively short

CHARACTER = 40 points
General body outline
Hump
Head
Horns
Coloration
Height of shoulders
Head hair
Beard
Cape & pantaloons on forelegs
Sex
  Males should be masculine, burly & massive
  Females should be more refined with a lighter shoulder & neck, and more refined head & bone
Hind quarters
Legs
Tail

QUALITY = 20 points
Well balanced with pleasing appearance.
Display alertness.

SOUNDNESS = 10 points
Correct body and skeletal structures
Free from any characteristics that would reduce the useful life of the animal
Feet
Legs
Eyes
Sex organs
The animal should stand and move freely without evidence of unsoundness

CONDITION = 10 points
Good health
Amount of fat
Amount of muscle

SIZE = 20 points
Big enough for age
(Bones & muscle development and body structure, not fat, best indicates size.)

calves, you don't know for sure if they can. Roy Houck advises buying youngsters, "as they are easier to handle and will adapt to their new surroundings and locate more readily."

Go to the sales. But don't bid until you've been to several and understand their strange tribal customs. Otherwise you may find that, when you picked your nose, you bought a herd you didn't want. The larger ranchers hold annual sales. So do state and national parks and refuges. The following list of sales was as complete as NBA could make it at press time:

Custer State Park, Hermosa, SD
Wichita Mountain Wildlife Refuge, Cache, OK
National Bison Range, Moiese, MT
Wood Buffalo National Park, Ft. Smith, NWT, Canada
Ft. Niobrara Wildlife Refuge, Valentine, NE
Sully's Hill Game Preserve, Ft. Totten, ND
Blue Mound State Park, Luverne, MN
Kansas Fish & Game Commission, Canton, KS
Shoshone-Bannock Tribes, Ft. Hall, ID

NBA also compiles and publishes results from the various public buffalo auctions. Almost without exception, buffalo sales are held in the fall, mostly October and November. The National Bison Range and Wood National Park sales are sealed bid sales. NBR opens bids in early September, WNP in December. The rest are on-site auctions where what you see is what you get. You take your choice and you pay your money.

LAND

7) How much land? Allow as many acres per head as cattle require in your area. This may be one acre per head in a high rainfall area to 100 acres per head in cactus country. Ask the advice of your county agricultural agent (look under county government listings: Cooperative Extension Service, Agricultural Extension Service, Extension Agent). Buffalo do not do well in a pen.

They like rough land to get some climbing exercises and some gullies and gulches and thickets for storm cover, since they will not enter barns or sheds. If nature didn't stud your land with boulders or thick butted trees or stumps for them to itch upon, set out some stout posts. This will please your animals and save your fences.

In this hungry world, it's a sin to waste good cropland on pasture and feed grains. Buffalo can turn rocks and brush into steaks and roasts, wool and leather.

Effective
Buffalo
Management

6
Handling
and
Facilities

THE GENERAL MANAGEMENT OF THE
AMERICAN BUFFALO

Encyclopedia Britannica, Inc.
Library Research Service

Buffaloes, like most other big game animals, thrive best where not closely confined. Large, open, grass pastures with a plentiful supply of clean, fresh water are best adapted to their needs. They apparently do not require shade from the intense heat of the midsummer sun, even in the southern parts of their range, but the presence of large trees against which they can rub adds greatly to their contentment.

Fences for confining buffaloes should be strongly constructed. For the outer boundaries of pastures and for smaller pens in which animals can be handled, six-foot, #9 gauge woven wire, supported by substantial posts, is recommended. Under normal conditions, buffaloes are not likely to test the fence very severely, but when excited, they may charge blindly into it, and then only one strongly built will withstand the impact. Fences made of five or six strands of wire on posts one rod apart have proved effective for interior paddocks when the animals are free from molestation.

For corrals and chutes, fences made of planks or poles are preferred, as the animals can readily see them and will avoid them, even when excited. Such fences also obscure the view of activities outside the corral and so reduce disturbance within, and finally, they are easy for a man to climb, which is a distinct advantage when he is seeking to escape an infuriated animal!

Barbed wire should not be used in buffalo fences. The barbs are ineffective in deterring the animals from attempting to escape and are a source of danger if they should become entangled in the wire.

33

FORAGE REQUIREMENTS

The foraging habits of buffaloes are, in many ways, similar to those of domestic cattle. Buffaloes graze principally on grasses, seemingly preferring them to weeds or shrubs. On southern ranges, buffalo grass (*Buchloe dactyloides*) and grama (*Bouteloua gracilis* and *B. hirsuta*) provide excellent pasturage. The shorter grasses appear to be preferred to the taller and coarser species. On northern ranges, in addition to buffalo grass and grama, wheatgrass (*Agropyron* sp.), bluegrass (*Poa* ssp.) and the smaller fescues (*Festuca* spp.) are readily eaten.

Buffaloes graze closely, and because of their habit of compact herding, they may rapidly deplete a pasture inadequate to their needs. It is estimated that two buffaloes require about as much pasture as three domestic cattle. Where conditions permit, rotating pastures will result in better utilization of the forage with less danger of depleting the cover and less likelihood of contaminating the ground with internal parasites.

In most parts of the United States, buffaloes may be ranged throughout the winter without supplemental feeding or artificial shelters. One herd, now numbering about 350 animals, is thriving in a wild state in the upper Tanana River Valley in interior Alaska. This demonstrates that cold weather and moderate snows are not detrimental. This herd was established from the nucleus of 23 animals introduced in 1928 from the National Bison Range, Montana.

When it is necessary to feed buffaloes during the winter or other period of insufficient forage, they readily consume hay and concentrated feeds. Alfalfa hay is palatable to them, as are other cultivated and native forage crops. Such grains as corn (preferably cracked or crushed), kafir oats, and barley, and cottonseed cake or meal are recommended supplemental feeds. Wheat is dangerous to feed because it expands in the stomach and may cause serious trouble. A maintenance ration will average 40 pounds of hay and six or seven pounds of cracked corn or five or six pounds of cottonseed cake an animal a day.

MINERAL REQUIREMENTS

To insure proper development, buffaloes need salt and other minerals in their diet. Most of these are ordinarily supplied in their forage. But on pasture soils deficient in essential minerals, the forage will likewise be lacking in them. One symptom of a common deficiency is a depraved appetite, resulting in the formation of such habits as "bone chewing." This deficiency can be relieved by supplemental feeding of steamed bonemeal and other mineral compounds prepared for domestic cattle. The salt requirement varies with the composition of the forage consumed, but generally buffaloes do not need as much salt as do domestic cattle.

Regardless of the extent of handling and the apparent domestication, buffaloes are wild animals of uncertain temperament and should not be trusted. Individuals have been trained for driving in harness, but in many instances supposedly tame animals have killed or seriously injured trainers caught off guard.

To move a herd from one pasture to another, some say men mounted on good horses are necessary. Yet many NBA ranchers would dispute this. They say horses and buffalo are ancestral enemies, that the buffalo hate horses and the horses are afraid of buffalo. Their herding choice is vehicles.

The following suggestions on driving were offered by W. E. Drummond: "When approaching a herd, the men should ride slowly, walking the horses, if possible, since that is the best way to keep buffaloes from running. When the riders are fairly close to the herd, they should stop completely. If they remain still, in most instances the buffaloes will approach to look them over. If the animals should start to run, some of the riders should circle in front of them, but at a considerable distance. Thus, by keeping the animals circling, they can be stopped. Once stopped, they should be permitted to quiet down for a while before the drive is resumed.

"When the drive is begun, one or two men should be stationed in front of the herd to keep the animals from running too rapidly. This is necessary, because if they run too fast, they string out and some will scatter, a few at a time on each side of the line, and get away. If there are not enough riders to slow up the leaders, the men behind the herd should push the animals at the rear toward the front and try to keep them bunched on the drive. They are not as hard to handle in a bunch as when strung out.

"When a herd is being driven around a hill or mountain, the leaders will always try to get away as they round the hill, and they will succeed if the rider on the outside does not ride at an angle or fast and hard. When an attempt is made to drive the buffaloes between two hills, they are sure to go around one of them if there are no riders in front. Riders must be alert in such a place in order to reach the brush ahead of the animals and cut them off.

"As buffaloes approach the end of a wing fence or a corral gate, they will try every trick they know to get away, and they know a number. In such a situation it is best to ride them fast and hard from the sides and rear until they have passed the end of the barrier, or are through the gate. Then they are easy to handle.

"There is always some cow in the herd that will come bounding out of the bunch toward a rider as if she would like to swallow both him and his horse. There is little danger from her, however, unless she has a calf, in which case it is a good policy to get out of her way. When the calf is young—a month old or a little older—if the rider will stand his ground, the cow will rarely charge his horse. If the rider runs from her, however, she will run him away from the herd every time he approaches. Then some of the other cows will join in the chase and keep the riders on the run all the time. A rope or a drover's whip will help put a stop to such behavior, but one should always bear in mind that no one knows what a buffalo will do and should govern his conduct accordingly. This much is certain: If a buffalo bull raises his tail and starts toward you, he must be given plenty of room!

"In small corrals, cows and young animals can be sorted and driven by men on foot, but with mature bulls or cows with young calves this is a dangerous practice. Workers should be constantly on the alert and be certain that an avenue of escape is always open.

"To handle individual animals for branding, vaccination, or other treatment, a properly constructed chute with a squeeze gate will facilitate the work and reduce the risk of injury to the animal. Young calves should never be roped around the neck with a tight running noose, as they are easily injured by such treatment. If it is necessary to rope them, they should be caught by the feet. If the rope is thrown around the neck, it should have a knot in it to prevent the noose from drawing too tightly.

"Disturbance of the cows during the spring should be avoided as there is risk of injuring them and causing abortion or premature birth of their calves.

"Individual buffaloes may be branded with a hot iron similar to the way in which domestic cattle are. If this is done in fall or winter, the wool should be carefully removed from the area to be branded, otherwise it may be ignited by the hot iron and the animal will suffer a serious burn."

GROWTH AND DEVELOPMENT

Normally, buffalo calves are born during April, May or June, although births have been recorded in all months. Newborn calves are reddish. This natal coat lasts until they are about three months old, when it is shed and replaced by the dark brown coat of the adult. The calves weigh between 30 and 40 pounds at birth. They grow rapidly and when one year old will weigh from 500 to 600 pounds. The cows continue to grow until six or seven years old, and the bulls until nine or ten years old. Extremely large bulls may weigh more than 2,000 pounds, but the average weight of mature bulls is about 1,600 pounds. The cows are much smaller, seldom weighing more than 1,000 pounds.

Buffalo cows ordinarily mate when they are two years old and bear their first calves at three. Instances of two-year-old heifers producing calves have been recorded, but are not common. Calves are usually born singly, cases of multiple births being rare. Cows 36 years old have been known to produce strong, well-developed calves.

Bulls reach active breeding age at about three years, but are not fully mature until eight years old. The breeding season normally occurs during July and August, but under artificial conditions induced by confinement and semidomestication, individuals may breed during all seasons. Mature bulls are difficult and dangerous to handle during the rutting period. The gestation period is between 270 and 280 days.

For butchering, buffalo one to four years old are usually selected, for young animals generally produce tender meat of better flavor than the older ones, although older, fat individuals may produce meat of high quality. Data are not available on the economy of feeding or on the best age at which to butcher to obtain the greatest financial return. It is believed, however, that butchering two- and three-year-old animals will yield the greatest profit to the producer. Buffaloes will dress about 500 pounds, and large bulls about 800 pounds. In adult animals, the front quarters are heavier than the hind quarters, but in young animals these are nearly equal in weight.

Records in zoological parks throughout the world, and observations of buffaloes on national wildlife refuges and private estates, indicate that the normal longevity of the buffalo is from 20 to 25 years. Some individuals have lived 30 or more years.

TIPS ON HANDLING AND FACILITIES

by Jim Hartmann, Everly, IA

The family and I operate 240 acres with 80 acres of pasture, 130 acres of row crop and the balance in hay and buildings. We liquidated our entire beef cow herd and increased our buffalo herd to around 50 head. For four years, I have been selling livestock handling equipment, and designing and installing corral systems. So when we got into the buffalo raising business, our first concern was for building the handling facility and fences we needed for raising buffalo.

Being in a service business, we knew there were two ways to learn what we needed to know to build our facilities. One being the trial and error method and the other being to ask those who have been raising them for years. We chose the latter and have learned much. And we are still learning.

Fencing: Your perimeter fences should be the best possible. The old saying "Hog tight, horse high and bull strong" fits buffalo, which have an inherited instinct to roam. Probably the ideal fencing would be either chainlink fencing or stockade (¼ inch rod feed lot) panels, both with barbedwire tops or higher in snow problem areas. These fences still may not be strong enough to withstand a head-on charge of a stampeding herd, but would be very discouraging to one animal with roaming fever. The fences as well would keep out the fearless, courageous, brave, bold, brainless, trespassing sightseers.

We personally prefer the less expensive strands of barbed wire with alternating wood and steel post with 16-foot spacing and a top barb height of 54 inches. This enables me to mow any grass growth in the fenceline thus reducing the snow buildup that often occurs in woven wire, chain-link or stockade panel fences.

Your pasture or range division fences need not be as strong as perimeter fences but should be strong enough and tall enough to detour any roamers. Three to five barbs 48 to 54 inches tall with all steel post with 10- to 12-inch spacing or alternating wood and steel post with 12- to 16-foot spacing are adequate.

Gates: Gates can range from the western barbed wire type to the modern tubular steel. One thing that I have noticed about buffalo, especially the bulls, they *do* know where the gates are even when they have never been through them. They have seen you go through and they know that they are the way out. From time to time they may even test the gate to see if they can get through. Cattleguards may also be used, but gates should be present, too, not only to allow passage of animals but also to close off the cattleguard should it become filled with snow. The gates in your feedlot and corral areas should be hung on hinges that will allow fast and easy swinging. They also should be equipped with latches that can be latched or unlatched quickly but that will remain latched even when tested by your roughest herd bull.

Feed Lot Fencing: This needs to be the best possible, because you are confining large numbers in small areas. Again, the types of fencing will vary from woven and barbedwire combinations to stockade panels to wood or combinations of all of these. Yet others may choose to construct fences like those in large commercial feed lots: made of oilwell pipe and steel rods. We prefer portable corral panels, either freestanding or attached to posts. The taller rodeo panels are best but the standard 64-inch panel will do just fine. Corral panels may be moved around, opened anywhere for gateway access, or the entire fence may be relocated without loss of, or damage to, costly fencing materials.

Feeding Facilities: For fattening bulls, some use self-feeders; others use in-yard feed bunks of wood, steel or concrete. Others choose fence line bunks. This is what we are planning in our operation as we continue to build our facilities. We are going to build our own precast bunk form and cast our own bunks. They will be similar in design to those for cattle but with a few changes. The bunk sections will be 12 feet long, sloped on both sides, not as tall on the side that the buffalo eat from and, instead of cable, we are going to construct a tubular steel fence over the bunks. This will prevent the hide damage that occurs when buffalo reach under a cable fence to eat.

Cow-calf operations may use similar feeding facilities during the winter when additional or supplementary feeding is necessary. For open range or field feeding, one could substitute some less expensive bunks of an old truck tire, one piece of plywood and an old tractor tire with the bead and side wall cut away down to the cleats on one side. But if supplementary feeding is necessary only a few days out of the year, ground feeding works well too.

Handling Facilities: This is as important a part of an operation as anything else and maybe even more, for without them it may be extremely difficult to market, care for, or sort your buffalo. Many things go into making a good working or handling facility. But before you start building, first get some ideas by looking at what others have, and see what may be commercially available (some cattle equipment may be adapted for use for buffalo). Determine what you need and want, decide where you want it located and then take a pencil, paper and ruler and build your corral on paper first. Changes are easier and less costly to make there than after you are actually building the corral! Generally speaking, there are several things to take into consideration in these decisions. To feed a cow-calf operation, you should have a corral system centrally located in the feeding area with easy access to all of the feedlots as well as to your winter pasture. Access to roadways for motorized transportation of your stock should also be considered.

Things that you may want to include in your corral system are adequate holding areas with lane ways or driving alleys leading to the corral. Feeding alleys alongside of fence line bunks work well for this. Your corral should have a sorting area where a two or three way sort is possible and holding pens to hold the sorted or treated animals.

In your alleyways leading to and in your corral itself, try to avoid sharp right angle turns as much as possible because buffalo being worked in a confined area easily become excited, especially calves. Buffalo moving down a driving alley may run blindly, not make the turn and collide with the fence where they can pile up, resulting in injury and even sometimes death. For similar reasons, avoid square corners in the holding pens. If the design of your corral will not accomplish this we advise that you simply build a short fence across the corners at a 45 degree angle. You may lose a little space but it may save you from some costly injuries to some very valuable stock.

Your crowding pen at the base of your working alley should be large enough to hold approximately six head. Larger pens may leave too much room and you may have trouble getting the buffalo to enter the working alley. The best design for a crowding pen is a half circle eight to ten feet in radius with a sweep gate or crowding gate that swings 270 degrees and follows the curved wall of the crowding pen and will hold or latch anywhere along the wall. The side of the crowding pen should be solid and approximately seven feet tall. A catwalk on the outside will give you a safe place from which to work. A dual exit design is ideal with one going into the working alley and the other into a load out facility.

Briefly, the load out should accommodate trucks, pickups and stock trailers. You could build an elaborate three-level loading chute—but a less expensive way is an adjustable height, portable loading chute for pickups and trucks. A step design for sure footing is best. For loading stock trailers, simply remove the loading chute and back the trailer in place. In most setups, the load out will be at a right angle to the working and driving alleys. To allow access from the catwalk to the working alley, a set of passage gates will have to be placed in the load out alley.

Your working alley should have several important features for safe and successful control of your buffalo. The alley should be basically solid sided, have a solid top, and solid rolling cutoff gates at both ends. The alley should be divided into sections with solid cutoff gates, leaving each section about 8 feet long. The alley should have a clear interior height of not less than 6 feet. The reason for the nearly all solid design is to limit the vision of the animals you are working. The less they can see, the quieter they will be and the easier they will be to handle.

Any opening should be covered with expanded metal. This will leave the interior of the alley smooth with no place for injury and will also reduce the possibility of horn loss that often occurs in handling. One side of the alley should be adjustable and with five settings from 18 to 33 inches. The non-adjustable side should swing open, allowing you to release an animal out the side of the alley without running it through the squeeze chute. The alley should not be more than three sections, approximately 24 feet long, not including the squeeze chute, if for no other reason than to save steps and time. One thing you may want to incorporate into your alley system is an individual animal scale.

Squeeze chutes are a very important part of your working facility. They can be a valuable time saver in working your stock. Like the alley, the squeeze chute should have a clear interior height and an interior length from tailgate to headgate. The sides of the chute should be adjustable in width and designed to squeeze from both sides to safely hold the animal from going down in the chute which can happen in a chute that squeezes only from one side. The chute should be equipped with a side exit or a squeeze linkage with a quick release that would allow one side of the chute to fall open in case an animal should go down.

The sides should have hinged bars and swing down for access to the upper part of the animal's body. The bars should swing down in pairs and the space between them should be covered with expanded metal. This will smooth the interior of the squeeze chute and reduce horn damage. It will also help prevent the animals from attempting to dive through the side openings instead of going straight ahead. The lower part of the sides should have a solid removable panel to allow access to the foot and underline of the animal. The floor should have cleats for sure footing. The tailgate should be solid and strong enough to withstand a pounding from either side. Likewise, the headgate needs to be extremely strong.

A stop or crash gate may be advisable for the front of the squeeze chute to stop the animal going through, thus allowing you time to close the headgate without missing. The stop gate should swing out of the way for easy access to the head as well as to allow you to exit the animal through the front of the chute. The chute, like the alley, should be covered with either a solid or expanded metal top to prevent animals from jumping out.

If the chute is manually operated, the handles should be long enough to give you the leverage you need. Short handles may be convenient to work around but may leave you short on leverage when you need to hold an animal securely in the chute, resulting in injury to either you, your buffalo or both. Also, avoid ropes as they can become tangled. It is easy to pull the wrong one, they often get in the way. When wet, they become slippery and are usually fastened to short handles providing you with poor leverage.

If the unit is hydraulically operated, the power unit, if mounted on the chute, should be mounted on rubber. But more ideally, the power unit should be located elsewhere to eliminate the noise that comes from the electric motor and hydraulic pump.

One may also consider placing a palpation unit between the chute and alley for access to the rear of the animal in the chute. The access gate should be hinged to the front of the unit which is opposite to the units used for beef. The gate should be a dutch door type, the lower part becoming a shield between you and the buffalo and the top half opening for access to him. The reason for the shield is that buffalo can kick fast, accurately, and viciously.

Transportation: Strength of the unit being used is important. Let's start with pickups. Avoid hauling buffalo in one with an open-sided and an open-top rack. If one is to be used, cover the sides and the top with canvas or plywood. Likewise, with farm stock trucks and fifth wheel trailers with combination boxes, the tops should be covered because buffalo will try to jump.

Another popular form of buffalo transport is the horse or livestock trailer. The more enclosed the trailer the better, because the darker it is inside, the quieter the buffalo will be and the better they will ride. If the trailer is dark inside it is wise to have a light in the front to aid in loading because buffalo are afraid of the dark.

Fiberglass trailer tops allow a great deal of light to enter and give the appearance of being open, which may cause problems. The buffalo may try to jump out and could damage the top doing so. But if you do have one already, you should paint the inside of the top black to reduce the light and you may want to line the inside of the top with a light expanded metal which could prevent damage should a buffalo attempt to jump out. Division gate or gates that can be hung on several of the trailers' vertical members change the sizes of the compartment for hauling mixed loads.

Semi-trailer trucks are also often used to haul buffalo.

Wooden floors give better footing than the all metal trailer. But if it is not possible to get them, bed the metal floor well to reduce the noise.

In the winter as well as in the summer, be sure to have plenty of ventilation in the trailer. Buffalo can stand the cold better than they can getting wet. Once you have them loaded, get them moving and keep them moving. Older animals left standing in a truck have been known to kick the side of the truck or trailer until they had crippled themselves.

General Handling: In loading as well as in handling buffalo, be patient, take your time and above all do *not* use an electric prod. The use of an electric prod will only excite, irritate and enrage a buffalo. It can even kill, because buffalo have a low tolerance for electrical shock. If you must use a club, whip or prod, a piece of 1-inch black plastic pipe about 4 feet long works best. It is safer for both you and the buffalo, it is light and easy to handle, it doesn't cost much and it is extremely tough.

If you are working buffalo on foot in a pen or yard, above all be careful! Buffalo are very quick and they are equipped with horns that can lay a man open like a sharp sword. Always be where you can reach safety in an instant. But at the same time, do not allow the buffalo to bluff you. If you run from them when they threaten, they will try you every time you work them. Never try to stop buffalo head on; you will probably lose. But you can deflect or turn them merely by standing off to one side and spreading your arms or by using a deflector of paper feed sacks or cardboard tacked to a piece of wood or by using a piece of thin plywood.

On the open range it is not wise to approach buffalo on foot. On horseback is better; but a vehicle such as a pickup is best. A mature bull can reach speeds of 40 mph and can maneuver like a jack rabbit.

In the winter, when the snow becomes too deep for wheeled vehicles, you may want to take to a snowmobile. Wild range herds may spook, but herds raised under more domestic conditions, used to noisy motored vehicles, probably won't be bothered by the snowmobile. Be sure your snowmobile can accelerate to 50 mph in seconds.

Shortly after we first got into the buffalo business, our herd of then just three jumped a neighbor's fence on New Year's Day and got out. Our first attempts to recapture them failed, teaching us several things.

First, we learned to be patient. Simply watch them go, because if they are not pursued they will stop sooner. But keep track of where they go.

Second, check the direction of the wind. You can't herd buffalo in any direction except into the wind. We did find that, during the calm of the night, we could maneuver the buffalo with a snowmobile by directing the headlight in the direction we wanted them to go. But you must keep them in the light, because if they get out of the light, they will stray rather than follow those they can see in the headlight's beam.

Buffalo also prefer going uphill to down.

If they ever catch you out on foot where you have no escape, no protection, play dead. You have nothing to lose. Generally, they are not aggressive but only defensive. If the person or thing no longer appears threatening, they will usually leave you or it alone.

If you have only the cover of a large rock, a tree or a tree stump, log or a shallow, narrow washout to take cover behind or in, do so! Even if you are not completely hidden, you may blend in well enough that the buffalo may lose sight of you. Their vision is somewhat limited despite the fact that they do have nearly 360 degree peripheral vision at close range. Your smell will give you away first.

I am no expert and I am still learning but I hope that some of my observations, experiences, and conversations with other buffalo raisers may be of help to some of you in working with and raising buffalo.

Wind-driven snow on the open flatlands creates special problems not experienced by those in protected, wooded areas. Drifts frequently cover fences; buffalo have walked right over and out and soon become accustomed to their new territory.

Many buffalo raisers spent the winter bringing their buffalo home time and again. Some finally stayed put, but others continued to wander even after the drifts were gone. Those unfortunate producers found themselves tragically and unwillingly in the meat business.

Shooting an entire herd seems a senseless waste as in many cases the problem could have been prevented or at least lessened had the owner done his homework and planned ahead. Vital to preventing such problems is preparing for the worst.

Snow will drift if there is something in the fenceline to break the wind. The solution is simple: remove the obstructions.

Most perimeter fences are either all barbed wire or a combination of barbed wire and woven wire. I prefer all barbed as it allows me to mow under the fence in the late fall to remove the vegetation. Some prefer woven wire as they fear the buffalo will crawl through the fence without it. If that is your preference, I would suggest placing the woven wire in the center with barbed wire top and bottom, keeping the bottom wire at least five to six inches off the ground (Illustration #1). An open bladed weed eater type mower can then clean up the unwanted vegetation on the fence line.

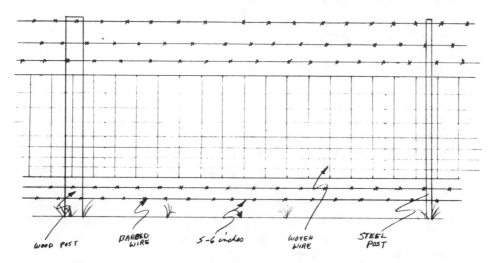

WOOD POST   BARBED WIRE   5-6 inches   WOVEN WIRE   STEEL POST

Mowing, to some, may sound like a lot of time consuming "busy work" but I've found it only takes me three or four hours to clear three miles of fence line and I have both types of fence.

There are alternatives to mowing; one is the use of chemicals that kill all plant life. The drawbacks to chemicals are erosion and, as the chemical effect wears off, the appearance of weed. Another alternative is fire. Burning off the fence line is less work than mowing, but it can also be hard on fences and shorten their life. I use controlled burning on part of my perimeter fence after I mow under the fence. The burn is to remove the growth in the road ditch and places where it is too rough to mow. This keeps the fence line free of snow and the ditch burning also keeps the road clear.

My buffalo are wintered on crop land, waterways, and hay fields with over two miles of fences. If, despite all my efforts, snow begins to build up on the fence, my first choice is to remove the snow. This is practical only if it is late in the winter and the problem is localized in a small area. Snow removal can be done with a front end loader or a snow blower. I prefer the blower as it's faster and it leaves no piles that may cause drifts and future problems later on. The snow blower pounds the snow into ice crystals which, when blown on the top of fresh snow, create a crust and prevent additional drifting.

My second alternate plan, the "if all else fails plan," is to move the entire herd onto a ten-acre hay field where I have only a little over a half mile of fence to keep clean.

This is where being prepared for the worst comes in. If you don't have a small field or pasture to use this way, a temporary one can be made by fencing off a corner of your present large field with a temporary fence. A fence of this sort can be constructed in the dead of winter without digging post holes or driving posts into the frozen ground. The wire is supported with posts constructed of one inch or 1¼ inch used pipe or old broken steel post four ½ inches long for uprights with a tripod base of three pieces of similar material 16–18 inches long welded to the bottom of the upright (Illustration #2). Your wire is held at the proper height by short pieces of ¼ or ⅜ inch rods bent into a cork screw design with an inside opening diameter of at least one inch. This permits fast and easy construction of the fence and allows the wire to move back and forth without tipping the post over. I place the posts 50–60 feet apart and use three barbed wires spaced 25, 40 and 55 inches off the ground. More wires may be used depending on the size area you are fencing and how well your buffalo respect fences. Some of the pressure can be kept off the temporary fence by keeping the herd well supplied with hay or corn stalks.

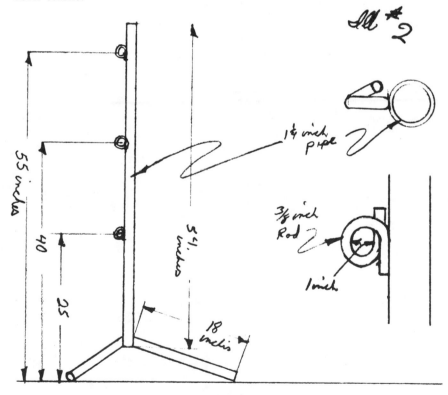

In feedlots and corrals, snow is always a problem. The more confined your buffalo are, the sturdier and more solid your fences need to be, and solid fences stop moving snow. I've found a couple of fencing types that work fairly well in these areas. One is a storm fence. This is a per-

45

manent fence which is constructed by placing posts 6–12 feet apart. To these you apply 2×4's or 2×6's horizontally, nailed two to three feet apart. Over them you nail either 1×12's spaced 11½ inches or corrugated roofing metal spaced two to three inches. All this is applied to the windward side of the fence and at least two rub boards should be applied to the other side (Illustration #3). I've found this type of storm fence with the spaced sheathing doesn't get a snow buildup right in the fence line itself and the buffalo also prefer the wind that passes through as opposed to the calm caused by a solid fence.

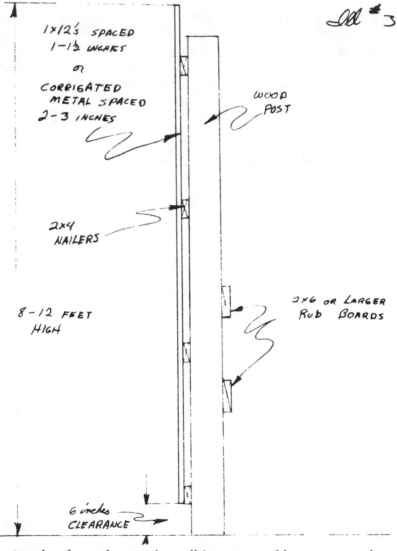

Ill #3

1×12's SPACED 1–1½ INCHES

or

CORRIGATED METAL SPACED 2–3 INCHES

WOOD POST

2×4 NAILERS

8–12 FEET HIGH

2×6 OR LARGER RUB BOARDS

6 inches CLEARANCE

Another fence that works well in snow problem areas can be made of portable cattle corral panels. They don't stop a great deal of snow.

But as the snow around them deepens you can simply dip them out and set them on top of the newly crushed snow. The portable panel may also be used to raise the height of a permanent fence buried in snow. Then come spring, with its mud, you can expand your feedlot to drier ground with panels. No posts are needed if the panels are set up in a curve or zig-zag pattern (Illustration #4).

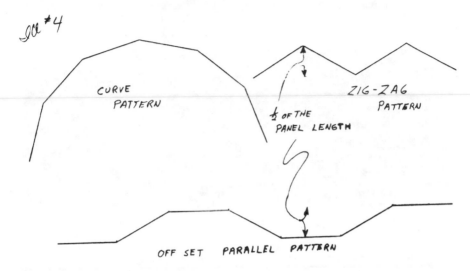

ILL #4

CURVE PATTERN

ZIG-ZAG PATTERN

½ OF THE PANEL LENGTH

OFF SET PARALLEL PATTERN

When raising buffalo in "snow country" it is a must to have some sort of snow removal equipment. Tractors with loaders are vital. A large skid-steer loader is another machine you may want to consider. If you like faster snow removal and want to avoid the mountains of piled snow, the tractor mounted snow blower is your answer. With this machine mounted on the right type of tractor, snow removal is a snap.

NBA MEMBER ENDORSES HIGH-TENSILE FENCING

by Bill Austin
Horse Cave, KY

We have been running about 20 head of buffalo, all ages, behind high-tensile (smooth) fencing since early June, 1981. Our fences were constructed according to the plans and specifications outlined in U.S. Steel's fencing manual for a 10 wire fence with the following exceptions:

1) 10 foot end assembly posts and 8 foot line posts were used.

2) An 11th wire was run 6 inches above the designed top wire making it a total of 47 inches off the ground. Our posts are long enough to run a 12th wire if needed which would make a total height of 53 inches.

3) We used Penta treated line posts rather than the recommended creosote.

Hi-tensile fence, made in New Zealand, has ratchets built right in.

And all end assembly, corner or other posts taking strain are oversize and very carefully set. If rock prevented our driving them to at least the 4-foot specified depth, then the holes were drilled, blasted, and the posts were set in concrete.

We have been very pleased with this form of fencing. My wife, Judy, who is from Australia, looks at it and feels at home. The young animals sometimes forget where it is and bounce off it while playing. There is nothing to hook a horn or scratch a hide—something that became very important to us recently when a blinded young heifer bounced off it many times.

When the calves were very young, they could pop through the fence if they pressed against it and, once they had learned this trick, would play a bit by changing sides. In all cases, a few strong words from Mum brought them right back to her side. This was never a problem and we've never had any of the larger animals on the wrong side (yet!). Once the calves grew a bit, they couldn't perform this trick any more either.

We left the 12th wire off in case we wanted to electrify it. Any of the other wires could be insulated and electrified as needed, but, so far, this hasn't been indicated. The far side of our 31-acre buffalo paddock is fenced by Federal spec wire mesh and a top strand of barb as it runs along about a quarter of a mile of Interstate 65, so you may assume we have a fairly calm little clan of buffalo here.

If you are seriously considering this type of fencing, please spend the

$5 and buy the *U.S. Steel Fencing Manual.* It may be the best investment you've ever made. It can be obtained from any dealer of U.S. Steel products. Check your yellow pages.

CATALINA ISLAND BUFFALO CORRALS

by A. Douglas Propst, President
Santa Catalina Island Conservancy

Like a lot of other buffalo ranchers, we started out trying to work our buffalo in cattle corrals with a cattle squeeze and loading chute. After a few near disastrous trials we modified a set of old cattle corrals for buffalo. This served as a learning experience and while the modified corrals were a vast improvement, they still left a lot to be desired. We began to visit other buffalo breeders and picked up a lot of good ideas, particularly from the fellows at the National Bison Range at Moiese, MT. We then selected a site at Catalina that could serve several purposes.

First of all, we didn't want to sacrifice any of our limited cultivated land but we did want the corrals adjacent to the fields so that they could be used as gathering areas after the crops had been harvested. We also needed a natural gathering area out of one of the main canyons, so we decided on a site in a narrow canyon which opened into one of the fields. This meant, however, the corrals had to be built in a long curve between the foot of the hills on one side and a creek on the other. We also had to do considerable grading and move the creek over. So far it has stayed in place.

A lot of input and discussion went into the design of the corrals, including suggestions from Dr. Dale F. Lott, animal behavior specialist at the University of California at Davis. Dr. Lott's theory is that buffalo attack each other in the corrals because they cannot, in tight captivity, go through their normal processes of giving submission signals. So, we designed our corrals to permit separation of the buffalo from each other quickly and easily. We also avoided corners wherever possible, but perhaps the main feature is that we do our sorting in the chutes, not in the alley.

Our work procedure is to hold the buffalo in one or several of the large pens in small groups until we are ready to work them. Then they are brought into the round corral which feeds a smaller round corral and the chutes. The chutes are equipped with sorting gates so an animal can be sorted into a tight corral or the alley for delivery to a holding pen. They can also be diverted to the scales and then sorted. A calf table and a full-size squeeze also work off of the chutes as well as a truck loading facility at the end. We separate the buffalo in the chute with counterbalanced blocking gates. These gates actually work on an eccentric and swing over behind the buffalo and lock into place. This again is a way to keep the buffalo separated and minimize injuries.

49

Corral-chute system feeds out of a canyon down to the creek. Curved to fit the land, it required considerable earth-altering and creek moving.

The squeeze was built in our shop by a very innovative cowboy welder, Bob Gaede. In fact, Bob designed and built all the gates and linkage systems in the corral. The squeeze has a box on the front of it so that an animal cannot jump through (lesson number one with buffalo is that they move at full speed and much faster than cattle). Once they stop with their head in the box, the squeeze is closed and the buffalo is restrained. Then the box is opened and there is plenty of room to work on his head, to take blood samples, check teeth or whatever. We also use nose tongs to hold their heads. This sure saves wear and tear on the veterinarian. The squeeze is equipped with movable swells inside to prevent lying down. This is particularly important during pregnancy testing.

The truck loading chute is designed with a quick acting gate so we can prevent buffalo from running back out of the truck. However, if one does turn around and gets by us, we have a gate at the bottom of the chute that will swing out part way to allow a turn around space. Once the animal turns, the gate is brought back against him and up the chute it goes. It should be explained here that all of our work in the chute area is done from the top ramps. This works very well, but perhaps most important, minimizes the amount of labor it takes to work the buffalo. Also, we should point out that if we need to prod a buffalo for any reason we put a few rocks in a beer can and tie it on the end of a long pole. It makes a great rattle and the buffalo move immediately.

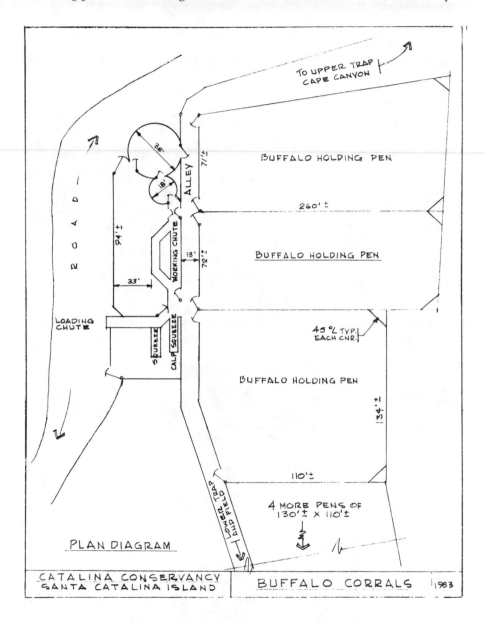

We have also learned the hard way that for some reason the buffalo always circle counter-clockwise in our round corral. Therefore, the sort gate in this corral is hinged on the right so that it opens into the traffic for easy sorting.

Bob Gaede with 3-position sort gate in the sort to the
right position. The same gate will also sort to the left
when operator pulls the handle toward himself.

Typical alley gate shows top and bottom hinge arrangement. Corral gates are the same type but covered with plywood to look solid.

Bufferfly gate latch on round corral gate. The whole mechanism is welded to steel quarter-round bolted and spiked to stout wooden posts.

Spring-loaded butterfly gate latch.

Top hinge with steel pin running down into the top of the hinge pipe.

53

Our gates are all made of oil well pipe and set on a pipe and collar cemented in the ground. They have a similar pipe running down into the top of the vertical hinge side pipe. This has an angle iron cap welded on it which is bolted with a long ¾-inch bolt to the gate post. Therefore, we have a no sag gate. They are comprised of a spring-loaded plate which only moves in so that when the gate hits the plate, it hinges in, allowing the end of the gate to pass and the plate then snaps back into a positive position and prevents any possibility of the gate being knocked back open. This is not only practical but a real safety feature.

We were fortunate in that, at the time we built our corrals, we had considerable salvage material available from a pier that was being demolished at Avalon. This provided pilings for posts and for corral lumber, bump boards, etc. In the pens, we used six foot, nine gauge chain link with a cable at the bottom, bump board in the middle, and a 4×6 plank along the top to make an overall height of seven feet. In the out traps, where there is little or no pressure from the animals, we used 11 gauge chain link with cables top and bottom.

The watering system is gravity flow from a spring. When we need power we use a portable generator.

We have had excellent success with these corrals and very few injuries to the buffalo. We normally load ten buffalo at a time on our truck, and this usually takes about ten minutes. Unfortunately, we do not have plans of these corrals available but hopefully the pictures will help. We will, however, be pleased to show the corrals to other buffalo producers at any time.

CUSTER STATE PARK IMPROVES BUFFALO CORRAL FACILITIES

by Thomas LeFaive
Administrative Director
Custer State Park

Custer State Park's buffalo herd is considered one of South Dakota's unique resources and one that Park personnel feel should be managed the way the people of South Dakota want a resource managed. Without the tonic of the wilderness, the buffalo life cycle would not possess the unique meaning that it does within the Park's boundaries.

With the melting of snow in April, mature cows instinctively separate from the herds. In groups of 15 to 25, they seek a high, secluded meadow in which to calve. Intruders to these private circles will find themselves quickly circled by the cows in response to their instinctive broodiness even before calving. This response is easily seen after calving until they regroup in large open grassland pastures. The calves may range from 40 to 70 pounds. Generally the male calves are heavier than the females.

The natal red hair turns to dark brown and black in three to four months and by the time they are six months old, they will weigh from 300 to 400 pounds with males averaging 20 pounds more than females. It is at this point that the annual Park roundup takes place and the calves are weaned, sorted and vaccinated for sale each November. Of the 400 animals sorted for sale each year, nearly 300 are calves.

LIVE BUFFALO AUCTION

The air is usually crisp on the third Saturday in November, when the annual Custer State Park Buffalo Auction occurs.

Four hundred head of surplus bison are auctioned at the live sale that attracts buyers from Oregon to New York, New Orleans to Canada. The demand for these animals makes them a worthwhile investment as hundreds of people now raise buffalo as a hobby and for profit.

The demand for buffalo on this continent far exceeds the supply, and prices paid are a premium when compared to beef. The bison, reported to have 25% more protein and about 25% less cholesterol compared to beef, is sought by many people for its delicious meat.

The high spirited nature of the bison makes the live auction that much more interesting, and the prime looking beasts bring many comments as they wheel about the sale ring. Though many sales are held throughout the continent, the Custer State Park sale may very well be the largest. With over 65% of the animals at the sale being calves, the average price for all classes runs around $400–$600. A prime two-year-old breeding bull may bring $1,300 and cow/calf units have brought as much as $1,500. The buffalo auction presents many people with an opportunity to obtain disease-free animals to raise, insuring the animals' continued existence. There is something unique about raising buffalo, something very American, possessing a spirit similar to that of the bald eagle.

The Park has enjoyed increased interest in its buffalo and its annual sale. This, together with the fact that the Park increased production from a little over 300 to nearly 500 surplus animals, justifies the need for good facilities.

Park Director Warren Jackson has sought approval from the South Dakota Department of Game, Fish and Parks Commission, for considerable improvement in the buffalo corrals where they work nearly 1,500 animals during three days of the annual fall roundup. In recognizing the need for additional sorting pens during the annual sale, the Park has nearly doubled the number of pens, permitting assignment of a pen to each buyer for proper distribution and handling of the bison once they are sold. The extra pens reduce stress as well and enable workers to sort and load much easier.

The working chutes, arena, auctioneer platform and catwalk to the platform have been reconstructed with steel. The Park has also sought

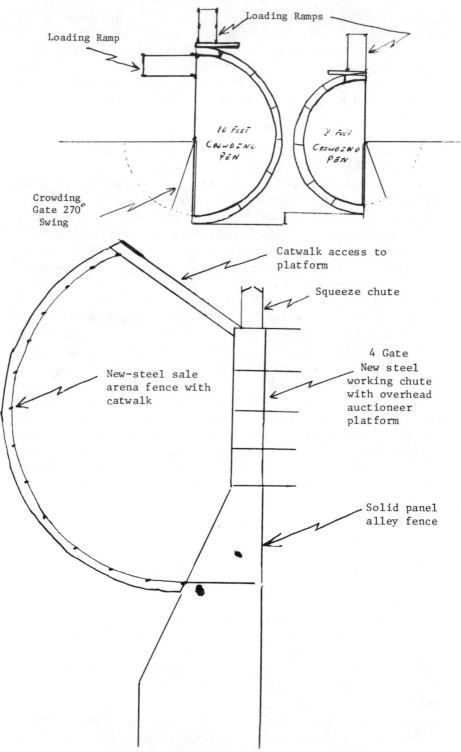

Loading Ramp

Loading Ramps

10 FEET CROWDING PEN

8 FEET CROWDING PEN

Crowding Gate 270° Swing

Catwalk access to platform

Squeeze chute

New-steel sale arena fence with catwalk

4 Gate New steel working chute with overhead auctioneer platform

Solid panel alley fence

Drawings of facilities at Custer State Park.

help in designing both an eight-foot and ten-foot half circle load-out pen. Jim Hartmann, Everly, Iowa, has spent much time assisting the Park with this work to have an efficient, safe and long-lasting load-out capability. Basic improvements consist of pipe, channel iron, and sheet steel mounted on I-beam frames that are supported with a concrete foundation.

Custer State Park believes the improvements are a worthwhile expense in terms of safety and efficiency but, most importantly, we seek the satisfaction gained by buyers who come to purchase the Park's stock.

An additional area of improvement is the funnel area leading up to the working chutes where horns have been broken off in recent sales. One-horn animals are less desirable and it is recognized as a problem which we hope we can reduce with the solid panels on the working alley fences.

The Park has gained much advice and some sound ideas from various people from around the country and we have put some of those to work to solve problems and make the working of nearly 1,500 head of buffalo more safe and efficient.

The sale starting time has been set at 10 a.m. to allow more time after the sale, because it is our opinion that loading-out at night, with or without lights, is an unsafe practice. The stress that is induced while loading at night is too great and results in severe losses.

We encourage buyers to come early and look over the new facilities and visit with Park personnel about the features. Park personnel will be available to visit with buffalo buyers and share information about management.

UNIQUE CORRAL AND SORTING SYSTEM

by Ray Smith
Longford, KS

Our corral was planned to perfection in every detail before construction began in August, 1980, at the Circle 3 Buffalo Ranch.

Including two large holding pens, the corral took over 2,100 pounds of welding rod and over 250 ton of pipe, angle iron, and sheet metal. The material used is four-inch pipe for the posts set in concrete 28 inches deep. The horizontal pipe is two inches inside. We used 95,000 pounds of 3/16-inch sheet iron from old steel grain tanks. Approximately 90% of the material was used.

There are features in this system never dreamed of before. The accompanying photos and diagram will explain it in more detail. The entire corral system is seven feet high. The crowding circle is 28 feet in diameter. The 1,000-pound gate, hinged on an eight-foot post, filled with concrete and rebar, has an electric motor on it and is controlled from the catwalk halfway up the side of the alley leading to the working chute.

Our fifty-foot-long alley divides into six individual sections. On the

RAY SMITH'S CIRCLE 3
BUFFALO RANCH CORRAL SYSTEM

Crowding Circle, 28' diameter
with 14' gate hinged on center post
to loading chutes, scales
and sale ring
location of
control box
for hex
gates
squeeze
Hex is 12' across with 6 gates
controlled from one location.
60'
68'
an 8' alley
circles the entire
corral

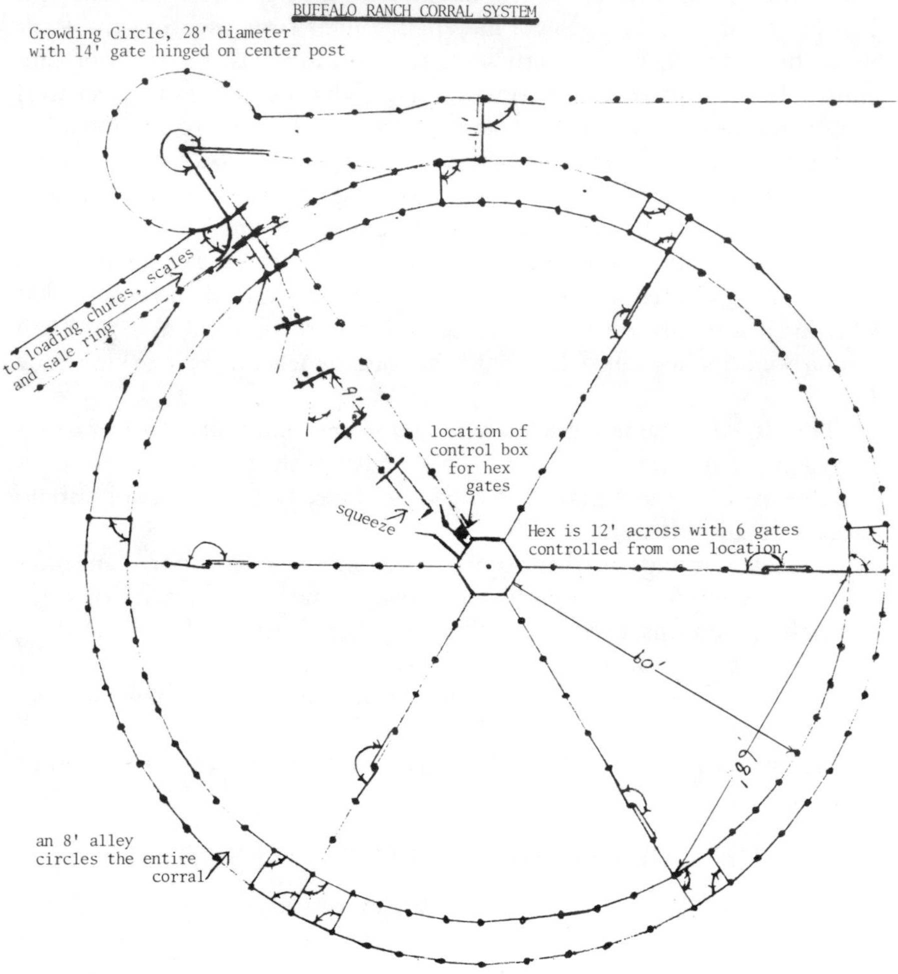

offside opposite the catwalk, the side of each section of alley is hinged. In an emergency, we simply pull a lever and the side swings open, releasing the animal.

Between sections we have two half gates controlled by pulling a lever from the catwalk.

When released from the working chute, the animal enters what we call the "hex." It is a six sided pen, twelve feet across. There are two half gates, 16 inches each, giving a 32-inch opening on each of the six sides of the hex. Each of the gates opens into a triangular pen 64 feet deep. The half gates are controlled from one location near the chute.

Only one animal goes into the hex, or sorting pen, at a time. We simply pull any of five levers to open the gate to any of the five pens. (The sixth opening is the entry.) The sixth pen is used as an emergency pen or can be filled from any section of the alley or the chute itself.

View of Ray Smith's corral under construction.

Center half of corral.   Eight-foot alley.

Crowding circle is 28 feet in diameter.

The system has been designed for safety for humans as well as the beast. There is a catwalk of expanded metal on all the division fences within the system and above the eight-foot alley which circles the triangular pens. A gate out of each pen enters the eight-foot alley, or we can go straight on out across the alley to the outside. There is also a gate between each of the six pens. Animals cannot be trapped anywhere in the corral. One can maneuver to any degree desired.

The eight-foot alley, the 28-foot crowding circle, and the alley leading to the chute and hex all have a concrete floor finished by setting a piece of expanded metal on the wet concrete, leaving diamond shaped imprints for traction.

The system is built on a slope for drainage. All the concrete has a slope to drain off in the natural direction of the hill.

All gates are hinged with 4½-inch pipe over the four-inch pipe posts. The three six-inch wide hinges on each post are supported by a two-inch band welded to the post. The latches are all automatic spring loaded.

We can exit from the crowding circle or the eight-foot alley or the emergency pen, sale barn, or the dual loading docks (truck or trailer).

Two men can efficiently work buffalo in this facility. One is on the catwalk and one at the chute. Sorting is simple. The operator of the chute just pulls a lever.

The system was planned, designed, constructed, and admired by Ray O. Smith and sons Warren and Robert.

The Smiths welcome visitors who soon also become admirers of the system. They will assist others in designing and constructing corral systems to fit their personal needs, circumstances, and available construction materials.

TO PUSH OR NOT TO PUSH

As we said in the preface, buffalo raising is not an exact science. What works for one operator, for one herd, may be wrong for another. We have a swivel chair opinion that it depends on the chemistry between the personalities of a certain herd and a certain operator. Change one or the other and you may have a whole new can of different worms. Here are two differing opinions:

Don't Push 'Em
You can't turn a buffalo. Don't push them too hard. Let them take their own pace and they'll go.—Gary London

Push 'Em
We turn our buffalo and we find we have a better time of handling them when we push them hard. We find when we let them set their own pace, they do go but every which way they want.—Jack Errington

ROUND-UP TIME!

by Jim Hartmann

Taking part in a buffalo round-up on the rolling prairie of South Dakota is probably one of the dreams of every buffalo enthusiast. I know it was one of mine ever since the tour during the 1978 fall convention of the National Buffalo Association, hosted by the Houck Family of Triple U Enterprises near Pierre, SD.

In the fall of 1979 that dream came true for my wife, Nancy, and me along with our two-year-old daughter, Penni. We spent seven short but wonderful, excitement-filled days on the Houck family's Standing Butte Ranch. Jerry and Lila asked us to come out and capture on film for ourselves and for the Houcks the 1979 fall buffalo round-up of cows and calves.

November 25, armed with plenty of warm clothes, two cameras, various lenses, fifteen rolls of film and loads of excitement, we all got into four-wheel-drive vehicles. Those steel horses may not be as picturesque as a beautiful Quarter Horse, but if you've ever chased or tried to outrun a buffalo, those horses of steel don't get near as tired and you don't get saddle sores, either. Sunrise found us heading northwest across the prairie toward the main herd with Jerry in the lead.

We soon came upon about 60 cows and their calves. The pickups circled around behind them and we began to move them to the south.

But they were not about to cooperate. As we attempted to bring them together and head them in toward the small holding pasture from which they are worked up into the main corral, they repeatedly broke off in pairs, keeping everyone going in every direction but the right one.

As less than half of them remained in the group, we regrouped our forces and set out to try our luck elsewhere. Our second attempt was much more successful. A group of around 120 cows, their calves and a few herd bulls were cut off from the main herd and headed in the right direction.

When bringing in a large group like that, Roy usually works as 'point man', the man at the head of the herd, whose job is to point the lead animals in the right direction and keep them headed that way. Jimmy and Lila work 'swing' which means that they swing the balance of the herd in behind the leaders and keep them together, trying to prevent any animals from breaking away. Sometimes a pair will break away and they (the 'swing') have the added job of trying to bring them back. Jerry usually pulls 'drag', he brings up the rear of the herd and keeps any slow pokes on the move.

Once the group is on the move, Roy heads them toward one of the gateways leading out of the main pasture. Roy chose to take them out the corner gate rather than trying to use a side gate because of the uncooperative nature of the buffalo that day. By going to the corner, we would have the advantage of the fences to work with. After they were through the gate, the herd was pushed on southward across another range pasture to the ranch buildings. The group moved quickly but restlessly along the fence to the ranch's main driveway. We crossed the drive and entered a small holding pasture of around 160 acres.

Once in the pasture, Lila dropped back to close the gate and wait, while Roy, Jerry and Jimmy cut off 25 or 30 cows and brought them back toward the gate. As they reached the corner where Lila is waiting they are turned and driven along the fence which runs parallel with the main driveway to our right. To our left, we pick up a wing fence.

As we come to the west end of the holding pasture, the fences form a funnel which forces the buffalo into a lane. This turns left and leads to the corral. Once in the lane, Lila turns right, crosses a cattle guard and goes around on the main driveway up to the corrals to make sure that the gate into the corral is still open at the bottom of the corral. Then she does her 100-yard dash up a slight grade to the next holding pen so she will be ready to close the gate when Jerry and Jimmy run the buffalo up into the second holding pen. I know why South Dakota ranch people are all so skinny. It's so far from here to there, that in order to get anything done you have to run.

In the lane leading toward the corral, the buffalo first head south about a quarter mile, where the lane turns right, heads west and up into the southeast corner of the corral. The lane is wide enough for at least three pickups to run side by side with room to spare. Once the buffalo are in the corral they are quickly followed in by Jerry and Jimmy. Roy stops in the lane and closes the gate behind them. After Roy secures the gate, Jerry and Jimmy work the buffalo up the triangular pen and into a smaller pen where Lila closes that gate behind them. From here they are moved into one of three other holding pens. Once penned, the gates are reopened and we return to the holding pasture by way of the main driveway. Then the process is repeated until all of the pens are filled.

As the pens are filled, any cows with late calves—"goldies" they're called because the calves are still gold in contrast to the older brown calves—are marked by Jerry with an orange paint ball shot from an air powered pistol. This can be a hit or miss proposition, for the paint ball never flies the same way twice, even though Jerry seldom misses. After the pair marking is done and the pens are filled, everyone takes position around the jenny barn.

Jerry cuts off five to eight cows with their calves, running them up the alleyway to the jenny barn pen. Once the last cow is in that pen, Lila latches the gate behind them. As the last of the group enters the pen, Jimmy attempts to keep them on the move along the south fence and into the short alley that leads into the jenny barn. But Monday wasn't Jimmy's day. Weather pressure systems weren't in favor of working buffalo. Some cooperated and some didn't. Those not cooperating sometimes sent Jimmy scrambling up a fence.

Once in the jenny barn, Frank, 81-years-old and the only non-family member residing on the ranch, closes the gate with a rope pull while Jimmy closes the roll-door. Frank used to ride and repair fences all summer but now takes care of maintenance and odd jobs around the ranch buildings. With some buffalo in the barn, Jimmy goes inside to run them into the outside pen. When some are in the working alley, the sorting begins. Operating cutoff gates were Frank, Kay (Jerry's sister who also lives on the ranch), Nancy and Lila. Jerry checks ear tags on the bulls for age and keeps a running count on all the animals going through, calling out "calf," "cow," "herd bull," "butcher bull," or "mark cow." Roy then pulls the rope on the appropriate gate to complete the four-way sort. Once all the animals in the corral were sorted, we headed back out to the small pasture to bring up more to work in the same way.

That afternoon, we emptied the small pasture and a second cut from the main herd was brought up and part of them worked. All things considered, the first day was successful with a little short of 200 head of cows being worked.

Tuesday morning we found some of the buffalo from the main herd had gotten out into the range (pasture) by the main gate. Those were the first ones we had to go after and we found them a bit more uncooperative than those worked the day before. After bringing those out of the main gate pasture and into the holding pasture, the day got back into a typical round-up routine of cutting off a group of 20–30 cows with their calves, bringing them up to the corral, penning them, filling the pens and then sorting. But with each draft that was brought in, they became more difficult to handle.

About midafternoon, a cow in the second draft coming up from the holding pasture, took the fence just as they came up to the entrance gate to the corral, broke through and took three calves with her. She managed to make her way back to the far end of the holding pasture, and in the process she went through two fences and took out three wire gates. We sorted those that were in the corral and spent the balance of the daylight hours fixing fence and gates. The day's total was slightly less than the first day's run but all things considered, it hadn't been all that bad.

That evening I decided to check on things at home and discovered things hadn't been the best at home either. The neighbor doing my chores had learned the hard way not to open a gate wider than a yard when the area was occupied with buffalo. While feeding some butcher bulls, one slipped past him and when he tried to run him back in, two more got out. They didn't run off too far, only half a mile through the corn field, up to my neighbor's yard. By the next morning they had found their way home and walked right in when the gate was opened for them.

On Wednesday morning we rose to find the sky to be overcast and there had been a slight skiff of snow. It was cold and the wind was blowing 30–35 mph with gusts up to 45. That day was wasted in every way. The buffalo were even hard to handle on the range. They continuously tried to break away from the group in the corrals, they wouldn't cooperate as you brought them up to a gate and would turn on you every chance they got. And in the jenny barn more than one was on the fight.

Over the noon hour the wind dropped, only to come back even stronger for the afternoon. Because of the dust and the fact that every hand was needed to hang onto the gates against the strong wind, we packed up the camera and pitched in to help. With the wind pushing 50 and gusts topping 65, it became almost impossible for even two people to open and close gates. By midafternoon and a couple of near injuries from wind-blown gates, Jerry called a halt to the operation. With a couple of daylight hours left to take advantage of, Jerry went out and brought in a load of large round bales of hay for feeding the calves which were already sorted off, held in part of the corral area.

Thursday morning found the wind down along with the temperature, but the sky was clear. I was surprised somewhat by the early morning sky on the prairie. The sky was full of stars that seem to shine brighter than they do in Iowa. Another thing that seemed so different was to see the horizon line as the first light of day starts to lighten the black

64

night sky. The horizon line was so distinct, the sky as it lightens the ground seemed to darken even blacker. The only thing spoiling the effect and giving a sign of civilization was one of those darn blue yard lights shining somewhere on the horizon.

We had a somewhat later than normal start on this day, as there were a few more chores than usual that morning. Roy, while checking some traps, also made rounds to some of the stock dams to break ice for those that didn't have flows (a siphon pipe set over the dam which provides running water for the stock at all times) already in operation. Jimmy and Jerry ground feed for the bulls on feed. After chores, we headed out.

There were only a few buffalo remaining in the small holding pasture. We worked them and then set out to bring up another draft. This first draft of the day was a bit difficult to get heading in the right direction. The group was mainly made up of bulls, a few dry cows and a few cows with calves. We worked that group quickly without much difficulty once we got them going. And then we returned to the main herd for still another cut. That second cut went like clock work, not like those of the first few days. By day's end, well over 300 head of cows had been run through. Not a bad day's run and with a late start at that.

Friday, with weather much improved, we headed out first thing to cut another drift off of the main herd. The buffalo worked the best they had all week. The cut was good sized and we spent all morning working them. After dinner, a second cut was made from the main herd. I braved it on foot that time.

About halfway out, they dropped me off to try to get some photo shots from a different vantage point. The view from atop a ridge near where they would pass was like something you can hardly explain. As the group came closer and closer toward me, I got this feeling of helplessness as I viewed them through the telephoto lens and I could feel and hear the hooves of over 300 head of buffalo pounding toward me. As they got to within 100 yards of me, they turned slightly and passed me by. There was a feeling of relief; but left also was a feeling of awe that I'm sure the early pioneers must have felt as even larger herds passed by their wagon trains. That day's total was again large with the operation running smoothly.

Saturday was for hunting. Three great white hunters from southern Wisconsin had come out to shoot trophy bulls. Everyone set out for the trophy bull pasture. Richard and Peter were the first to shoot their pick of the bulls. It is not hard to see why it was so easy for the 19th century hide hunters to destroy the herds of buffalo. From a safe vantage point (the back end of a pickup) the first tried his luck with a black powder rifle. But, as forewarned by Jerry, the shot was unsuccessful in downing the large bull. One might say he managed to give the bull only an ear ache. Jimmy is the back up man and he quickly dismounted and, from a short distance from the pickup, delivered a single fatal shot which dropped the large bull at once. Quickly, Jerry bled the bull. He was winched onto a trailer and then the second hunter tried his luck. Convinced that a more modern weapon would be most effective, he picked out a bull standing some distance off among some others and downed him with a single shot.

After this bull was bled and loaded, we returned to the ranch slaughterhouse where the bulls are unloaded and the Houck family members put on their rubber knee boots and aprons to start butchering.

Roy winches the bull inside off of the roll-in table that is just outside the building. Quickly, the tail and hind legs are skinned out. The bull is then lifted and the hide is split down the belly, starting just below the tail and ending between the front legs. The front legs are then skinned out and Jimmy proceeds to cape-skin the head and shoulders. First he splits the hide from atop the hump. Then he hand skins the neck and front of the shoulders.

When the neck and shoulder hide is loose, Roy fastens a second hoist around the horns and begins to pull the skull out of the hide. As Roy lifts, Jimmy peels the hide off with his razor sharp knife.

After the skull is removed, the hoist is then tied around the hide of the front legs and the front legs of the carcass are then tied to a ring imbedded in the concrete floor. The hide is simply pulled off. This pulling is faster and easier than using a knife and also prevents nicks in the hide that often result when using a hand or air knife for skinning.

After the hide was off, Jimmy gutted the carcass and Jerry split and hung it on the carcass rail. As Jerry and Jimmy are finishing up on the first hide, Roy starts in on the second and Lila takes the skinned out hide outside and spreads it out on the ground to salt it down with 25 pounds of salt. The skinned carcass is then weighed, taped and placed in the cooler.

After lunch, it's Harold's turn to bag his trophy. He boasted he was going to down his bull with a pistol. Jerry attempted to discourage him from trying, as handgun shooting has shown to be even less successful in downing buffalo than black powder shooting. Harold was determined to be the first so he took aim with the Smith & Wesson .44 magnum. The shot roared, the bull's head swung up and to the side and he began to walk and then trot off, but went less than 300 feet. Jimmy was getting into position to down the big bull with the rifle and remarked that another hunter had tried and failed with a handgun when the bull tumbled to the ground. The downed bull was the first that anyone had managed to drop on the Standing Butte Ranch with a handgun. Excited, Harold jumped out of the pickup to deliver a second (insurance) shot. So excited was he, he ran around behind the bull as we shouted "shoot him again!" He kept yelling, "where, where?" and Jimmy repeated, "in the head, in the head." The hand held cannon roared again, but it was not necessary as the first shot had definitely done the job. Again the bull was bled out and photographs taken to chronicle this first handgun shot buffalo.

After loading the hides and heads of the trophy bulls and some frozen meat, the hunters departed for their homes in Wisconsin.

By this time it was late afternoon with less than two hours of daylight left. Jerry said he was going out to check the flows on the stock dams and check to see how scattered the remaining cows and calves were. This meant it was flying time. We rolled open the doors to the airplane hangar and there sat the ranch's Piper Super Cub, a single engine, high wing, cloth covered, two seater, tail dragger. We rolled it outside, Jerry turned the engine over a couple times by hand and we climbed aboard. A few quick turns of the engine with the starter and she sprang to life. The sod runway on Standing Butte runs northwest to southwest, but what little wind there was was out of the east. So instead of using the runway, we simply ran down the driveway and within a couple hundred feet we were airborne.

We swung west over the cows we had sorted off earlier in the week and could see the butcher bulls and yearlings that Jimmy was taking feed out to. As we passed over a stock dam, Jerry would swoop down like an eagle to check the water flow and then climb just as rapidly back to our cruising altitude. As we progressed northward the view of the Oahe Reservoir was breathtaking in the setting sun. Near the north end of the ranch we found the remaining cows and calves. They weren't too wide spread. We also found that a semi-truck loaded with large, round hay bales had lost the center out of his load as he had rounded a corner. This sent bales careening into and through the fence, much to the delight of the buffalo.

Jerry and I returned to the ranch and put the airplane away. Then we drove back out to repair the fence and round-up the buffalo that had gotten out. When we reached the broken fence only one bull was still out and as we drove up he came back in on his own. It was past sundown by this time and quickly grew dark. We repaired the fence to the best of our ability in the pickup's headlights and returned to the house for supper.

Sunday morning instead of going cross-country after the last buffalo, it was quicker to go around on the road and enter the range pasture on the northeast side where we had fixed fence the night before. The air was crisp and the bulls were in a playful mood. By the time we had the entire group pushed together, we saw that it was by far the largest number we had tried to bring up all week. At first Jerry was a bit concerned that it might be too big, but these buffalo were the best behaved of all.

With little testing, the herd moved slowly south. By the time we had the first of this herd going into the holding pasture, the tail of the herd was nearly a mile back. We managed to get the holding pens in the corral filled and worked a few of them by noon.

Unfortunately, after we finished eating, Nancy, Penni and I had to head for home in Iowa. Our planned four-day visit had already turned into seven and I still had a crop to finish harvesting at home.

In all those days of working nearly 2,000 head of buffalo, only three animals injured themselves enough to draw blood. The Houcks handled their buffalo, some of the biggest and finest breeding stock around, in the gentlest way you could possibly handle a wild animal.

If any of you ever have the opportunity to be part of the Standing Butte fall round-up, don't pass it up. It is worth every minute you spend there.

Those were the most wonderful seven days I ever spent—not only capturing in pictures and memories some of a buffalo round-up, but also working with some of the world's best buffalo handlers and learning from them.

7
Feeding
Procedures

Generally speaking, feed that's good for cattle is good for buffalo. They eat mostly the same grass, hay, and grains, with a few exceptions.

Buffalo eat more kinds of things—they'll eat some of the weeds and brush that cattle won't. This is why you can often pasture three buffalo on an acreage that will carry two cows. On the other hand, buffalo avoid loco weed and other cattle killers.

Acorns are toxic and wilted wild cherry leaves will kill buffalo the same as cattle. We have no reports of stunted sorghum or sorghum regrowth killing buffalo as they kill cattle, deer, and sheep, but unless you have a buffalo herd to spare, we suggest you keep them away. Since buffalo have the same stomach arrangement as cattle, we fear they, too, might be subject to cyanide poisoning from sorghums. Sorghums include such crops as kaffir, sorgo, milo, sudan grass, johnson grass, hygeria (high-gear). It's the drought-stunted plant that kills, and any fast growing plants such as young shoots or fast regrowth following cutting or frost.

Also avoid pine needles as too many can cause abortions, as in cattle. Alfalfa is too laxative for buffalo. Fortunately, they won't go ape over the lush young legumes that tempt cattle to overeat and bloat.

Buffalo prefer good, honest buffalo grass or prairie hay. Some ranchers feed alfalfa hay cubed with grains. As you would expect, they need more variety than one kind of grass or hay. They prefer the finer grasses, such as buffalo grass.

Adult buffalo will usually refuse grain unless they were trained to it young. Grain is unnecessary for adult buffalo except as a finishing feed for slaughter. Certainly you'll have fewer problems feeding buffalo than in feeding cattle: Less going off feed, less scouring, rarely (if ever) problems with bloat and founder.

Buffalos' wild instinct will guide them, if you'll let it. Give them feed ingredients and minerals free choice . . . and some still, small voice down deep in their innards will tell them precisely how much to eat of which to balance intake to need—more accurately than a whole bank of computers.

SCOURS

Under natural, or range, conditions buffalo do not develop scours (excessive looseness of the bowels). They suffer from this condition only when closely penned, subjected to massive re-infestation by intestinal parasite eggs or larvae. Ranchers with plenty of land avoid this by rotating pastures.

"Internal parasites lay eggs on grass," relates Roy Houck. "By changing grass, your buffalo won't eat grass with eggs on it. We hold some pastures a year, some two years, so that the buffalo go in on clean grass. Parasites don't get a chance to get a hold of them that way."

Tapeworms and roundworms can be a real threat. A modest worm infestation can drain an animal of a pint of blood a day! If you can't rotate, give a worm bolus annually. Chemical control is moving so fast, it's a good idea to check with your veterinarian. Until specifics are worked out for buffalo, go the cattle, horse, or sheep route.

BLOAT & FOUNDER

Some ranchers flat out say buffalo won't overeat, bloat, or founder. Others hedge a bit and say they don't ordinarily, but they might if forced to gustatory indiscretions by unfamiliar feed, confinement, or other artificial conditions. As pointed out in the later discussion of storm survival, more cattle die from overeating, founder, and bloat after a blizzard than die from the blizzard itself! Buffalo have the wild animal's instinct to avoid this problem. If an animal dies quickly and swells up, the post-mortem finding may be 'bloat'—but is more likely clostridium, which is explained in a later chapter.

PASTURE

Buffalo with plenty of pasture and fresh water usually appreciate a good home and show no desire to roam. Tempted to build nice snug barns for them? Don't! They won't go in. Natural breaks and thickets is all the shelter they will accept.

Sow the grasses adapted to your area (your County Agent can suggest some); give preference to those offering high protein, especially varieties like the bluestems that stand up through the snow and retain their protein content in the dead stalks. Go easy on the high-carbon species such as Timothy.

In the South, buffalo prefer the short grasses such as buffalo (*Buchloe dactyloides*) and grama (*Bouteloua gracillis* and *B. hirsuta*). On Northern ranges, they like the tall grasses such as wheatgrass (*Argrophyron* spp.), bluestem (*Andropogon* spp.), bluegrass (*Poa* spp.), and the smaller fescues (*Festuca* spp.).

Requisite acres per head depends mostly on rainfall; the figure is also affected by age, size, and condition of your livestock, productivity of

Buffalograss

BUCHLOE DACTYLOIDES
TYPE: perennial, dioecious, stoloniferous, dense sod former.

HEIGHT: short

STAMINATE HEADS: (male) raised above leaves.

PISILLATE HEADS: (female) appearing as if a bur in the axils of the leaves.

BLADE: curly, sparsely hairy on both surfaces.

DISTRIBUTION: primarily in southeastern Montana, usually on heavy valley soils and in overgrazed drainage ways. It is dominant over large areas of the Great Plains.

DISCUSSION: Buffalograss spreads rapidly by stolons to form a dense matter tuft. It withstands heavy grazing better than any other grass species native to its region. The foliage is very palatable and nutritious. It cures well and furnishes excellent winter feed. Many reports of buffalograss have been blue grama. Both species are very poor forage producers.

the land, forage varieties, etc. A good rule is to stock your buffalo at the same rate as cattle in the area. Because buffalo are more efficient grazers, you give yourself a built-in cushion.

ROTATE YOUR PASTURES

Pasture rotation reduces parasite re-infestation, promotes grass vigor and production and makes more efficient forage utilization. The secret of mini-management is to allow enough acreage so that plenty of the sun ripened grass will stand above the snow to carry through the winter full of pep and vinegar.

The Triple U Ranch keeps some pastures ungrazed for winter feeding.

A more intensive management method is to rotate pasture thus: Save winter pasture; graze them all summer in one pasture, leaving another untouched. Then when the snow blows, open the gate and let them drift into the reserved forage, close it behind them when they're all through (drifting may take some days). Hold the herd there late enough into the spring to let your summer pasture get a good head start. Some ranchers with lots of land keep three or four seasonal pastures. And some reserve a pasture for hay.

Another approach is to use smaller pastures, rotate several times a season. This works best in a high rainfall area and requires worming.

PRAIRIE GRASSES

by Clinton S. Fraley
Executive Officer
Clay County Conservation Board
Spencer, IA

Throughout the United States we have three major prairie types. From east to west these would be the tall, mixed, and short grass prairies.

72

Between the eastern border of the Great Plains and the Pacific Ocean we had nearly 800 million acres of "prairie." Needless to say this was the natural range of the buffalo before man left his footprints on the land. Now, it could be safely stated, the tall grass prairie is extinct east of the Mississippi River plus the states of Missouri, Iowa, and Minnesota.

The mixed grass and short grass prairies still survive west of the Missouri River and are utilized by our western ranchers. This area currently provides nearly 100% of America's open rangeland—as opposed to pastures. Is it productive? YES!

Prairies must be studied from east to west, to get the complete picture. Rainfall, soil quality, fertility, and production decrease as one travels from east to west. Thus, if one is interested in establishing *warm season grasses* (native prairie grasses) they would have to locate their own operation on this east-west line to determine the type of grass most favorable to their location.

When writing about native grasses, it is impossible to state which species will grow best and provide maximum protein in a given area. Rainfall, fertility, altitude, soil pH, etc., must be identified before species selection occurs. Therefore, it would be best to seek a local contact to help provide this information.

Native grasses did provide enough protein to sustain the bison in the past and could do so again on a more limited basis. A pasture rotation of warm season grasses (native) and cool season grasses (domestic) could provide season-long grazing. If one is interested in establishing native grasses, the following would be just a few to consider by major prairie type.

The tall grasses would be Switchgrass, Big Bluestem, and Indian Grass. These are the most productive, and for grazing a single species, would be the simplest to establish and maintain. Across the mixed grass range, Little Bluestem and Western Wheatgrass would be good choices. Farther west in the short grass country, one would try to establish Buffalo Grass and Blue Grama.

The above species are just a few of the most productive for each prairie type. Local conditions would dictate what species would actually be planted. These grasses may be established east or west of where they naturally occur, such as planting Buffalo Grass east in the mixed grass areas of Little Bluestem west on the short grass prairie areas.

To obtain help in making selections on the local level the Soil Conservation Service maintains an office in almost every county in the U.S. They have studies on the most productive grasses for the county along with very good information on local soils and soil fertility. The S.C.S. has access to all types of technical information and they want to share it with their local landowners. Seek their help when you have questions on prairies.

BEWARE OF CONTAMINATION

Beware of what your neighbors may be spraying in adjoining fields. One buffalo producer lost 28 head to the chemical Aatrex. Aatrex is commonly used to control pests in corn fields. His neighbor aerially sprayed an adjoining field, and the chemical drifted over his buffalo pasture, contaminating the grass.

Ingesting a heavy dose can kill a buffalo in two or three days. With a lighter dose, they may linger for weeks or months. The major symptom is scouring. Since death may occur long after the initial Aatrex contamination, and since scouring can be a symptom of other problems, always have an autopsy performed.

The product is an herbicide used primarily, if not exclusively, on corn. It is a product that stays in the soil up to three years, preventing growth of grasses, weeds, etc. For some reason it doesn't affect corn. The lethal properties diminish in short order—usually in eight or 10 days it ceases to be poisonous if ingested. The label recommends a 30-day "cooling off period," however.

Since most corn is treated with this stuff, and since corn is commonly used as silage for livestock, it either loses its potency or it is not harmful to common barnyard-type critters. It may be an isolated quirk, dangerous only to buffalo. The animals in question were posted and the diagnosis was conclusive.

MANAGING BUFFALO ON PASTURE

Supplemental feeding may be necessary in drought to prevent overgrazing and damage to the grass. It also helps the animals hold their gains. If you're not striving for top gains and if there's surplus pasture to preclude overgrazing, the buffalo can get along on dry grass and wet water just dandy.

Some ranchers feed ear corn directly on the ground to supplement winter pasture, figuring two to three pounds a day. Pellets or range cubes also work well. If the snow gets too deep, you may have to blade off an area. A bonus benefit is that you can coax the herd through a gate and into another pasture or even into the corral by laying a trail of goodies they're used to. One rancher buys stale bread for little or nothing. He reports buffalo like this treat just as children like candy. He can coax them anywhere with a trail of stale bread.

Ranchers without excess land buy hay for winter feeding. Buffalo will consume less hay than cattle—about half a bale a day—because they are neater in their habits; they trample and waste less. The prudent rancher will keep a year's supply of hay stacked up against an unusually severe winter. If you do this, and it all rots while your bison merrily burrow through the snow for grazing, rejoice. Consider it an insurance policy you didn't need.

74

Like all wise livestock producers, the Triple U lays up
some feed in fat years.

Buffalo root in the snow for forage like hogs, while cattle starve. Even if the snow is armored in ice, you'll see buffalo crack the ice with their fore-hooves and root out huge slabs. They'll crack the ice on ponds or streams to get to water, while cattle in the same pasture won't even have sense enough to drink from the holes the buffalo make! I've seen an ice storm coat every blade of grass with ice. The buffalo feasted. The cattle bawled in starving self-pity.

FLUSHING

Graziers for centuries have known that 'flushing' both males and females with grain or high protein supplement six to eight weeks before breeding insures a higher conception rate and lustier young.

Supplemental feeding for six weeks prior to calving is reported to produce stronger babies, more milk for them, and ready breeding-back.

SALT

Keep salt available at all times. Some operators prefer a mineral salt. Some find a fly repellent in the salt works well. Some mix wormer in the salt.

WATER

If you don't have enough watering places on pasture, livestock will overgraze near water, while lush forage goes to waste at a distance. Buffalo will walk farther from water to grazing and back than cattle will. You will get more even grazing and more efficient forage utilization, while animals walk off less of their grain, by providing water on one-mile centers so that no animal need walk more than one-half mile to drink. Actually, buffalo can make it three or four miles from grass to water and back, but you'll get uneven, wasteful grazing if you force them to an endurance test.

SCRATCHING AND PARASITE CONTROL ON PASTURE

Buffalo are fond of rubbing and scratching. In the days of the old west, they are reported to have destroyed army barracks with their rubbing and scratching. So-called buffalo wallows were actually dust basins where the buffalo rolled, choosing alkali soil, to get rid of cooties and old hair. Naturally the rain pooled in them and turned them to mud. Soldiers favored them for ready-made foxholes during fire fights.

If your pasture lacks boulders, stumps or stout trees for rubbing, better set out some strong posts for them to rub on, and save your fences.

Western ranchers don't need oilers (sometimes buffalo just plain don't like these). Buffalo roll and dust themselves in alkali spots. If you don't have alkali soil, try super-stout cattle oilers.

There's a spot on the flank and loin that the buffalo can't reach with either tail or tongue. Here flies eat them raw. This is where the salt containing fly repellent works best. Cattle dips and sprays don't work very well. Besides the labor, there is the danger and the damage of trying to push buffalo through a spray or dip tank, for which they have no desire!

WEANING AND FEEDING CALVES ON PASTURE

One weaning method is to separate the calves and turn them into pasture where you've left round bales on the ground and a self-feeder of grain. Also self-feed protein supplements through the winter.

Small "domesticated" herds may give good results with this creep-feeding. A creep-feeder is a feedbunk in which grain or other supplement is available to calves at all times. Low or narrow entrances keep adult animals out, while permitting calves to enter and leave at will. When situated in the open, as a buffalo feeder must be, it is roofed to protect the feed. Large, extensively managed herds may not adapt to creep-feeding. It may help to include some older calves that know the ropes; they will encourage the youngsters to enter the creep and eat.

Besides teaching them to eat grain and acquainting them with bunks and enclosures, creep-feeding spares the pastures and produces faster gains. It predisposes the calves to make good feedlot gains later, and preconditions them to follow bait trails for easier handling.

Ranchers recommend that at this stage you accustom them to a chunky bait like range cube that is easier for animals to pick up with less waste. They're more likely to find cube shapes in the snow rather than kernels or small pellets.

"When feeding in a bunk, I usually use a small pellet," says Kenneth Throlson, D.V.M. "But when feeding on the ground, I use a cube about ¾ inches across and from 1½ to 3 inches long. Even small calves can pick up one of these and chew it without waste. It seems that they enjoy the larger cube and consume it more readily. For one thing, by picking it up and lifting the head to chew, they are in a less vulnerable position for being hooked or bunted. With small pellets, they must keep their head and eyes directed down to consume the same amount of feed."

In the wild state, or even if left to themselves on your range, mothers would be inclined to suckle their young two or three years, letting the brand new calf starve. This results in a 50% or poorer calf crop. You can up this to 85% or 90% by weaning every fall. Confine them in a pasture (ideally) well away from their mothers, and out of hearing. Otherwise the anxious mothers may wreck your fences to rescue their darlings. Buffalo calves tend to bawl less when weaned than domestic calves. Creep-fed calves wean better with less noise and no stunting because they start right in on the grass without having to be starved to it.

One rancher who drylot finishes hundreds of buffalo bull calves every year advises "weaning and feeding calves when they are eight or nine months old. This enables the cows to winter better and produce a stronger calf each year. The biggest benefit is training the calves by handling during the feeding period. This gentles the animals and teaches them to eat all kinds of feed, a big benefit when the animals are put on full feed for slaughter." Such calfhood preconditioning will result in their going onto a full feed of a hot ration from day one in drylot.

If you have spring, summer, and fall calves, this means three separate round-ups, three separate bunches of calves to wean, handle, and feed: Another advantage for bunching your calving into as short a period as you and your bulls can manage.

NO PILLS AND POTIONS NEEDED

Pill peddlers swoop down upon new owners of small herds who haven't had much experience, convincing them of the need for all kinds of vaccinations, injections, vitamins, supplements, and medications, warn experienced ranchers. When buffalo can roam free in a nearly natural environment, on land that has not been leached or cropped to death, they select their feed and take care of their own needs. Only buffalo closely confined or grazing deficient soil will need any bottled protection. You're dealing with a wild animal. The more you leave him alone, the better he gets along.

FEEDLOT FINISHING

Some people prefer grass fat buffalo. However, the urban American taste, long educated to cornfed beef, finds grain finished buffalo the supreme victual. We have this advice from experienced buffalo feeders:

"Though many advocates of buffalo meat insist grass fed is equal in quality to grain fed, grain feeding definitely has its advantages in certain seasons. A 2½-year-old buffalo slaughtered in the fall while its weight cycle is on the ascendancy will produce a juicy morsel you'll find difficult to distinguish from grain fed meat. If you wait a couple of months and slaughter that same animal after he's endured harsh winter weather and poor feed, while his weight is on the decline, you'll find the quality of meat far inferior to grain fed. For maximum meat quality, always slaughter while the animal is gaining, whether on grass or grain."

"Finishing ratios will vary with what you have available," says Dr. Thorlson. "I buy all my feed and this is what I use since it is the cheapest and does a good job for me:

"I keep them on long hay all the time when feeding them. I start them on a few pounds of the pellets the size they make for sheep and in about 10–14 days have them on full feed:

screenings	1,497 lb.
corn	333 lb.
vitamins A-D	20 lb.
mineral (1:1 Ca:P)	50 lb.
molasses	100 lb.

"Some nutrition major may groan at this, but it does a good job for me and I'm buying a finished pellet for less than $90 a ton. With pellets, there isn't a dust problem and they are easily eaten with no waste and the molasses makes it very palatable.

"Vets and ranchers always need to be concerned about the ratio of calcium to phosphorus. Usually, for ruminants your ratio will run from 1:1 (equal parts) to 2:1 (two parts calcium to one part phosphorus). Most probably run around 1.6–1.8:1. I like a 1:1 since any legume hay is already high in calcium."

Select your surplus bulls two to four years old and put them in a feedlot with super strong fence, gates, and bunks. Feedbunks must be brawny and anchored to the earth. Otherwise a bored bull will entertain himself by tossing the bunk about. Bunks suitable for cattle are too frail and too high for buffalo, whose short, thick neck makes it tough for them to raise their heads shoulder high. About 20 inches is the best bunk height. If it's higher, buffalo can't reach over it comfortably; if it's lower, they climb in.

FEEDLOT RATIONS

Hard grains like corn, small grains, and sorghums should be cracked

78

for more thorough digestion and less waste because grain is not ruminated by older animals. Calves chew it thoroughly. You don't need to crack it for them until they become old enough that undigested grain begins showing in their droppings.

Along with grain, you need to feed protein supplements such as cottonseed cake, soybean oil meal or cake. Give plenty of prairie hay. Go easy on antibiotics and growth stimulants. The consumer is so mistrustful of stibesterol and other hormones that smart buffalo raisers are scrupulously avoiding artificial growth stimulants and most injectables, thereby appealing to the health food market.

Sun cured alfalfa meal pelleted with cornchop, molasses, cottonseed meal, plus vitamins and minerals especially formulated for buffalo gives outstanding results.

You can expect two- to four-year-old buffalo bulls to eat about two pounds concentrate and 20 pounds of prairie hay per day. Other rations are offered by NBA ranchers. Check the index for "rations."

One rancher's favorite ration for drylot is mostly corn and a protein supplement with plenty of hay free choice, 60 to 90 days. These animals are mainly bulls about 30 months old and weighing 1,000 to 1,200 pounds when put on feed. They will weigh from 1,200 to 1,500 pounds when ready for slaughter, dressing out at 600 to 800 pounds, choice grade.

HOW LONG ON FEED?

Most ranchers agree a 90-day feed is best for putting a prime finish on buffalo and producing tasty meat. Others find it profitable to go as long as 140 days. Feeding is best done during cool months; buffalo don't do well in the lot during hot weather. Buffalo have starved in commercial lots because they didn't know what a feedbunk was. This is another reason for introducing calves to grain in a feeder. At about 90 days, most buffalo either stop gaining on the hot ration or gain so little it's no longer profitable.

Even a 60-day finish will put three to four pounds a day on them. They gain 25 to 50% faster than you can expect on cattle, and on less feed. Finish feeding makes the meat more tender, juicy and flavorsome and, while buffalo have no wild or gamey flavor, more pleasing to the cultured taste.

A two-comin'-three-year-old bull will weigh into the feedlot at about 1,200 pounds. After 60 days on full feed he'll weigh out at least 1,400 and dress about 58%. Bulls three-comin'-four, will give you up to 900 pounds dressed meat.

Unlike cattle, buffalo need no warm-up in the lot. You can self-feed a hot ration the first day, relying on their wild instinct to govern their own intake and avoid scours, bloat, founder and all the other digestive ills cowflesh is heir to if given full feed too fast.

SOME IMPORTANT DONT'S

Don't mix young, small buffalo with older, bigger animals because the big ones will horn the young ones away from the feed.

Don't crowd your buffalo. They need twice as many square feet per head, twice the linear feet of feedbunk, compared to cattle. If they're too crowded, they'll fight too much and not do well. They can kill each other. Packing houses report massive internal injuries, even with unbroken skin, from goring. Some producers dehorn to avoid this loss.

Don't expect a buffalo to thrive on feed that is incapable of making blood, bone, and meat. Life is a series of delicate chemical reactions, and when one chemical is missing from the ration, life may continue, but the growth, condition, and finish will not occur. For top production, animals must have a high energy, balanced, and complete ration.

Don't put buffalo into a commercial feedlot unless the operator has a record of success with buffalo. You don't want to pay for his training. Even though cattle feeding experience puts you 'way ahead in the buffalo feeding business, you still have lots to learn.

Don't skimp on rations for mother cows. If you build a puny foundation on a calf because of inadequate maternal nourishment, you'll pay three times as much trying to rebuild a two- or three-year-old to market finish.

DURHAM BULLS ON FEED

by Jack Errington
Assistant Manager
Durham Meat Company

Surplus bulls, culls, etc., from the Durham Bar D Ranch herd at Gillette, WY, are finished in custom feedlots—one at Powell, WY, the others in California. The following is based on an interview with Gary London, office manager at the Alta Verde Industries' commercial feedlot at Powell.

The animals are started on baled hay with 10% dry ration on top of the hay for five or six days until they get accustomed to the feed trucks. Some buffalo, like some cattle, don't care for grain at first and will pick out the hay, so periodically we hold the hay and make them clean up the grain.

The basic ration consists of corn silage, alfalfa hay cubes, and a small amount of cracked corn—about one pound per day, plus a protein supplement. Precise ingredients depend on local availability and relative costs at the time. In general, a good cattle ration is a good buffalo ration. You can start buffalo directly on a hot ration as long as they have plenty of hay free-choice; just don't try to give buffalo an all concentrate ration. Buffalo will eat about 2¼% of their body weight a day, compared to cattles' 2½%.

80

You can't predict what buffalo are going to do. They may not go as hot on the ration as cattle. The procedure is basically the same but progresses faster than for cattle.

Sturdy gates are quick and easy to manage.

45° corner panels prevent bunching-up in corners, reduce damage to animals and to fences.

Detail of simple, strong latch.

Feedbunk headrail height is
adjustable to size of animals.

Double panel division fences
reduce fighting between
lots.

MANAGEMENT IN THE LOT

The pens pictured hold 150 head each, although Alta Verde has some
500 head pens. You put over 150 head of buffalo in a pen and confusion
results when the pecking order breaks down. Alta Verde allows two feet
of bunk per head of horned animal; otherwise the smaller animals, those
lower in the pecking order, get horned aside and go hungry.

All night yard lights add to security and help prevent spooking the
buffalo. Also, they eat all night and make faster gains.

8
Breeding
for
Maximum Profits

Buffalo breeding, like feeding, is easy. Just let them do what comes naturally. You can interfere by helping too much—or you can encourage them in directions that improve on nature.

The first requirement is plenty of room. Large bunches of animals confined to a small lot do not breed dependably. You can't put a bull and a cow in a 15×20 foot pen and expect romance to blossom. They just don't have enough room to maneuver and court.

We frequently read that, in the wild, buffalo cows calved every other year. Modern ranchers doubt this. They think probably most cows calved nearly every year, but favored the older sibling over the younger and let the wee one starve. We'll show you here how to wean a 90% or better calf crop, utilizing modern ranchers' experience.

Nature provides a bull calf for every heifer, plus a few to spare. In the wild, the dominant bulls established their harems and drove off the competition until age caught up with them. On the ranch, the best bull to cow ratio comes out to 1:10 in big herds; while in small herds, 2:10 or 2:15 insures a better calf crop.

TWO YEAR OLDS ARE "IFFY"

A two-year-old bull is mature enough to breed, but some ranchers report his callow advances are rejected by mature cows who know about the birds and the bees already. The older lovelies seem to prefer bulls in the three+ age bracket. Many, given the choice of a two or nothing, choose celibacy! This can cause you a lot of disappointment come spring when no little red bundles from Up Yonder arrive.

However, two-year-old heifers get along beautifully with two-year-old bulls. The Durham Bar D Ranch has run what they call a co-ed herd: They kept their two-year-old bulls and heifers together, 1:1, separate from

83

the mature breeding herd. The calf crop from the co-ed herd hit 95%, the highest they ever got. This practice also saves wear and tear on two-year-old bulls who would otherwise try to fight the big boys and get whupped.

Buffalo come to puberty a year later than domestic cattle. The buffalo heifer usually mates at age two and drops her first calf at three, although a two-year-old mother is sometimes seen. Bulls become dependable breeders at three and remain productive well into their 20's, although ranchers find it profitable to retire a bull to the meat factory while still in his prime. Practice varies—some turn their bulls at four, others wait as long as nine years. The older they get, the bigger, stronger, meaner, more dangerous, and harder to handle they become. A four-year-old is still good eating. After that, he begins to give all buffalo flesh a bad name.

Whereas a domestic cow has outlived her profitability at age 10 or 12, buffalo cows routinely reproduce into their 20's and have been known to drop a calf after 40. Bar D retires its cows at nine to help insure reproductive dependability.

MATING IS A SUMMERTIME THING—ON THE RANGE

Mating begins slowly in mid June and builds to a crashing climax in late July on the central Plains. The rut occurs progressively later northward, peaking in mid August in northern Canada. Thus Mother Nature brings most of her calves into the world late enough to miss the worst spring blizzards.

Estrus (heat)—the time during which the egg is capable of fertilization, lasts about two days. If the cow does not conceive, she comes back into heat in about three weeks. If she is impregnated, estrus does not recur for another twelve months, during which time she will have birthed her calf.

The bull forms a tending relationship with his chosen cow that endures for a day or two. Zoologists call this a "temporary monogamous mateship." During this time, he runs off, if he can, any other buffalo that comes too near whether cow, calf, or bull (he's also likely to run you off, too, so this is a good time to stay away). If a bigger bull displaces him, he seeks another mate. This temporary monogamy explains why you need two bulls for 10 cows in a small herd. You have a spare if the other one is busy when another cow comes into heat.

Pawing, horning the earth, and wallowing are all part of the rutting pattern. Bulls usually limit their competition to a simple poll-to-poll shoving contest, with the loser backing down. Battles to the death are less common. Fatalities are more common in confined herds because the loser can't get away.

On the ranch, under confined conditions, breeding may occur at any season, ranchers report. Buffalo in the wild, or maintained under nearly natural conditions, do not need to be separated by sex, as they will automatically do what comes naturally at the proper time.

On the other hand, buffalo under crowded conditions will begin to cycle like cattle. If you are finding many out of season calves, you would be wise to keep your bulls away from the cows except in mid and late summer. A cow who produces a calf in November won't drop one in May.

The other alternative is to keep her separate from the bulls after she's calved late, and hope she enters the estrus at the proper time the following summer. Perhaps your veterinarian could precipitate her estrus at the time you want with a hormone injection. Ask him.

And remember, for a good calf crop you need good cow nutrition. Roy Houck advises, "Be sure your cows are in good healthy condition; otherwise they will not come in heat. Every time we've had a drought our calf crop has been down 8%–10% the next year."

It's a big advantage to bunch the calves. If calves are all born within a few weeks, that means you round-up, brand, vaccinate, wean, dehorn, etc., only once a year, instead of two or three times. This saves labor, expense, and the risk of repeated round-ups. Too, calves of uniform age and size do better at the feedbunk; otherwise, the big ones hog it.

Cows bred in July and August will calve in April and May, avoiding the worst blizzards and giving calves enough summer growth to make use of fall pasture. Gestation is the same as the human and bovine: 280–285 days, or just over nine months.

FLUSHING

Graziers for centuries have known that enriching the animals' diets with higher protein feed about six weeks before breeding improves reproduction rates. Bar D does this for their cows. Putting bulls on a grain ration for about the same period should help them do a good season's work, too.

SEPARATE OUT THE BULL?

Some ranchers advise keeping bulls and cows separate except during breeding season. This may be good advice on small spreads, especially if the animals are a bit crowded.

The Standing Butte, which has more room than most, finds this unnecessary. "We have never kept our bulls separate from our cows and 90% of our calves are dropped from April 15 to June 1," Roy Houck reports. "Nature has a pretty good way of taking care of this time factor."

CALF BE NIMBLE, CALF BE QUICK

Buffalo calves are born a cinnamon red and weigh about 40 pounds. Multiple births are extremely rare. The newborn, as soon as he gets some milk, is a marvel of mobility and endurance, capable of following mom, even in a blizzard. A domestic calf born in a blizzard usually is abandoned by its mother and freezes.

The natal red grows darker, until by summer's end, the calves are almost as dark as mom 'n pop, and weigh 100 to 200 pounds. A yearling calf should top 650, a 2½-year-old bull ready for drylot feeding should weigh 1,200 to 1,400 pounds and put on 150 to 200 pounds in 60 days of full feed.

Bulls are full grown at eight, when they weigh around a ton, give or take some. The biggest buffalo of record was Bar D's eight-year-old "Jumbo," who topped the scales at 3,340 pounds in his bare feet in 1969.

GENETICS

Little is known of buffalo genetics. As are all wild animals, they are highly homozygous (genetically similar), a characteristic doubtless enhanced by their near extinction. Most of the U.S. herd today stems from the few hundred left alive in 1900. They all trace their ancestry back to these few progenitors a relatively few generations back, and therefore carry little genetic variation. This is proved in the virtually identical colors. Differences in size, horn curvature, etc., are thought by breeders to be essentially environmental. The Rocky Mountain herds, for example, have been observed within living memory to change slightly to a form more adapted to their environment—an effect thought to be due strictly to environment. Take them back to the rolling plains and they'd most likely revert to the Plains type in a few generations.

Stormont Laboratories of Woodland, CA, working with the Department of Interior, has blood typed two entire herds. Some basic genetic differences have been found in these separate herds. Both have been closed herds since the early 1900's. Until more entire herds are blood typed, little will be known of the genetic variations in buffalo.

If buffalo maintained a family tree, some could trace their ancestry back to hybridization experiments 40 or more generations ago. The likelihood of your buffalo producing a calf with domestic cattle traits are remote, however. Buffalo-cattle crosses are not a species; they are a hybrid. The species a hybrid breeds with will predominate in exhibited traits, and in several generations the adulterant species will be diluted beyond recognition. At this point, blood typing will indicate domestic blood in your buffalo's genetic history, but it can't tell how much there is or when hybridization occurred.

When a buffalo carrying an identifiable cattle gene breeds with another carrying this gene, there is a possibility of a calf with domestic characteristics. If that calf breeds with a certified pure buffalo, the offspring will be a buffalo, although it may still carry the domestic gene. Somewhere down the road (nobody knows for sure how many generations it takes), the cattle gene disappears entirely and you once more have 100%

pure buffalo. For the most part, this is what we have in this country today: pure Plains bison.

The idea of combining the vigor and survivability of the buffalo with the docility of domestic cattle is centuries old. Many have tried it. The first result is terribly high fetal and maternal mortality. The next result is sterile hybrid bulls.

The consensus of nearly all who have tried it is DON'T. The Canadian government tried it for 50 years at Manyberries and Wainwright, then chucked it. If you must give it a whirl, be sure your are financially and emotionally prepared for horrendous losses.

If you are in doubt about the purity of your buffalo, or simply want to satisfy your curiosity and verify the fact, NBA will help you.

BE CAREFUL ABOUT INTRODUCING "NEW BLOOD"

Be careful about buying buffalo out of a zoo or any confinement. "When we were building our herd," reports Roy Houck, "I went into Missouri and bought some. They were in such poor shape we had to tail them up out of the truck. But by spring you couldn't tell them from our own.

"All that was the matter with them was malnutrition. I was lucky I didn't introduce some disease into our herd—I wouldn't buy any like that again. We don't buy any from outside now. We have a healthy herd so why jeopardize it?!"

ALBINOS

The likelihood of your buffalo producing a white calf is even more remote than a throwback from old hybridization experiments. Here again, a family tree would help. Back in the days when millions of buffalo roamed the plains, it was estimated that albinos occurred once in about 10 million births. How accurate that estimate is, no one knows. But it was rare indeed. Albinos were so rare as to be held in religious awe by the Indians, who valued an albino buffalo skin beyond all else, for its strong medicine.

We also know that the albino gene survived the slaughter and was carried by at least a few of the thousand or so buffalo that remained at the turn of the century. We know this because a white calf was born on the National Bison Range in Montana in the 1930's. He was entirely white except for a brown topknot and lived to age 26. His offspring were all brown. (Technically, a true albino will have no color whatsoever: white hair, pink skin, eyes, hooves, and horns.)

In the 1960's a white calf was born on the Standing Butte Ranch in South Dakota, but did not survive. In recent years, calves with white patches have been reported. Whether these are the result of albino genes is unknown. They have all been crippled to some degree and have not survived. The white hair may have been an expression of another mutation.

For an albino to be produced, both parents must carry the recessive albino gene. Then the chance of their producing an albino is one in four. There is no way to identify a recessive gene, so you can't breed for this trait until an albino shows up. With no natural intermingling of the species, the likelihood of two critters with the necessary recessive genes getting together are slim to none. Until another albino is born, we won't know if the recessive albino gene has survived modern culling.

INBREEDING AS A TOOL

"Inbred" is usually pronounced with a sneer denoting decadence and inferiority. Yet inbreeding can be the buffalo raiser's most powerful tool. Like dynamite, it is a power for good or evil, according to how it is used.

Inbreeding is the mating of close relatives. Brother-sister mating is the closest inbreeding possible with mammals; then parent-offspring mating. Linebreeding is the mating of less close-in relatives.

Knowledgeable breeders recognize inbreeding as the most effective and rapid way to "fix" type and isolate "pure" (homozygous) lines. Most wild populations are highly homozygous as can be seen by their similar markings. All buffalo are about the same color, for example.

What inbreeding does is increase genetic purity whether for good or ill. It concentrates recessive as well as dominant genes, giving recessive characteristics a chance to show up. Recessive characteristics are usually less desirable than dominant traits.

A few ranchers are linebreeding and report favorable results from the practice. Too little data is in, however, for NBA to endorse the practice. Stormont cautions that linebreeding or inbreeding can be dangerous to the financial health of your buffalo, as linebred buffalo with little genetic variation to begin with can swiftly succumb to the problems that plague linebred cattle. Usually a small calf crop is one of the many problems you may encounter with this practice.

"Inbreeding," says the *Encyclopedia Britannica*, "does not create defective offspring. It is not harmful in itself." Nor does inbreeding create genes. ". . . inbreeding usually increases the proportion of defective, weak, slow growing or otherwise undesirable individuals." Again, it concentrates genes, or, as *Britannica* says more scientifically, the genes are ". . . brought together in combinations that would be rarer under outbreeding."

In one classic experiment, laboratory mice were inbred for many generations but were rigorously culled and only the *top* individuals of each litter saved. This strain became bigger, healthier, producing bigger litters, etc. So the secret of inbreeding success is merciless culling.

The Little Buffalo Ranch, Gillette, WY, started in 1922 with two cows

"Big Medicine," the white—but not albino—buffalo born on the National Bison Range in May of 1933. He died at the age of 26.

and one bull and has never introduced any new blood. The herd today is an excellent one. In harvesting the surplus bulls, the practice is to take out those that are "not as pretty" as the best bulls. This is discussed more in detail in a later chapter.

While few scientific studies have come to light, a knowledge of genetic principles of mammals suggests that close mating among buffalo (father/daughter, mother/son, brother/sister) does not carry as much power for good or ill as does the same practice among cattle, because the genetic sameness gives fewer differences to work with.

The reason for the apparent limitation on buffalo selection, as hinted above, is that they do not carry the rich mixture of widely differing genes that cattle, even of the same bloodlines, carry. It is a principle of genetics that, the more widely the parents differ, the more hybrid vigor the offspring will display. Of course, when the parents are of species that differ too widely, either the offspring will be sterile (witness the mule), or conception will not occur.

Until genetic studies have been made and the genes mapped out on the chromosomes, which takes many buffalo generations (and, so far as we know, is not being done), the breeder's best advice is to mate the best to the best and eat the rest. Selecting for size, vigor, feed conversion efficiency, reproduction rates, weaning weights, conformation, meatiness, etc. (Review the Bison Judging Criteria in Chapter 5.)

In practical ranching, you'll breed any cow capable of dropping a calf and top out your bulls to serve them. Until someone takes the time and expense and trouble to isolate breeding units, to make detailed studies of the physical characteristics, to keep precise records, and to production test his bulls, buffalo genetics is going to remain a misty field of guess, gosh, and try anything that works!

WHAT DOES A GOOD BUFFALO LOOK LIKE?

Nobody has ever spelt out exactly what we're looking for in a good buffalo. Do we want long, rangy bodies—or short, compact frames with a big hump? Do we want light color or dark? A few breeders are just beginning to work along these lines, developing criteria on exactly what characteristics make up the ideal buffalo.

Also developing is a registration program. This is a herdbook, which will require certification of purity first and then single sire breeding units and detailed parentage records for subsequent steps toward registration and pedigree. Raising registered buffalo is not the goal of most who enter the industry, nor is it feasible for many who do not have the facilities to maintain single sire herds, or the desire to keep detailed records. Most are satisfied to just raise buffalo. But even those folks practice stringent culling and strive to get the best from what they've got.

CULLING FOR GROWTH AND VIGOR

by Dave Raynolds

Current culling practices among buffalo raisers appear to have two conspicuous effects. First, the total number of buffalo in private herds has been growing for more than fifty years. (In contrast, the smaller total number of tax supported herds—federal, state, and local—seems to have stabilized.) Second, the health of privately owned animals improves over time from the care they get. This is probably true also for those tax supported herds which receive active management.

What are the current culling practices which ensure that more and healthier animals survive? This addresses questions in the minds of those with small and medium sized herds, and readers with no buffalo at all. It is drawn from experiences related by raisers of various sized herds, but the comments may not apply to the very largest operations. (From National Buffalo Association membership figures, it appears that more than 90% of raisers have fewer than 50 animals.)

"Keep everything" is the principle upon which most buffalo raisers start. Beginners with a handful of animals aren't planning to dispose of any at first. They worry about fences, accidents, and nameless diseases conjured up by neighbors (who don't have buffalo). They're intent on maintaining what they have and increasing the herd size. The natural strength of the animal works in their favor most of the time.

After a while, though, something unexpected may happen. "In our case," reports Dave Raynolds, "with thirteen in the starter herd, Livia, the lead cow, had a premature calf at seven months which died in a March snowstorm. Eight months later she still seemed to be understrength. It took us that long to make our first cull decision. Meanwhile, we had six fine calves and the prospect of buying more heifers. We killed Livia for meat and found (1) no sign of any disease on autopsy, but (2) a second, twin calf had died when about a five-month fetus, had mummified, and was only partially absorbed. Meanwhile, Livia had not gotten pregnant in the 'open' horn of the uterus. In the vet's opinion, she might not have conceived for several years, and the mummified fetus might have caused her future health problems while it was being absorbed.

"So after 'keep everything' comes 'improve the herd.' We shared some meat from that first kill and found ourselves terribly popular in the area. People would stop us on the road and ask, 'when are you going to kill the next buffalo?' We began to think about that, though we had still not reached optimum herd size. While part of the brain still said 'keep everything,' there was the case of Cleo, a pink tongued cow of nasty disposition that charged riders on horseback. We were running low on meat, so a year later Cleo went 'for safety's sake.'

"That second cull is probably the one that crystallizes the pattern most appropriate for the individual herd's operation. Thinking ahead to your optimum herd size, you begin to plan how many calves you'd like to have each year. You look at your cow-bull ratio. You look at the age of your animals. You consider the problem of handling them if you plan live sales. As you near optimum size, you plan to select 'keepers' including perhaps a few for insurance against accident or 'natural' loss in an optimum sized herd.

"For example, one large meat producing herd regularly harvests an age group, in their case 2½-year-olds of both sexes. Older cows and bulls are maintained. The cows are scheduled to live 'forever,' while sometimes an old massive bull is sold to a trophy hunter. From time to time a 'replacement heifer' makes it past the 2½-year cutoff, usually by calving early. This herd had maintained this natural rhythm for more than a decade at its present size.

"In our case, the eventual problem was going to be the cow-bull ratio. Land and feed availability suggested that our optimum size might be about 50 animals of one year or more. If split evenly by sex, there would be 25 females and 25 males, an inefficient ratio in terms of the resultant calf crop. A ratio of 38 females to 12 males would be better. On the male side, that would provide the herd bull and two eventual replacements, with nine male weanings available for sale as long yearlings when 1½. Assuming the calf crop was evenly divided by sex, there would be nine female weanlings, about that number of two-year-old heifers (of which one or two might calve as an exception), about seven three-year-olds due to calve, and 12 older productive cows. The group ought to produce about 20 calves each year. We could see that, to maximize production quickly, we had to select for cows and against bulls.

"Our third cull choice was the result of an accident. Since we're basically running a breeding/live sale operation, and aiming for size, we want the bulls to look super. Herman, a very strong young bull, also seemed sexually advanced for his age, and was flirting with cows as a two-year-old. Maybe Monster, the herd bull, nicked him in summer pasture, or maybe he rammed a rock. Anyway, Herman lost 1½ inches off the tip of one horn. This meant we couldn't sell him off to someone to be their herd bull, since he'd look less than perfect. Though his genes were fine, we had better 'keepers.' Herman made fine eating.

"Our fourth cull was another young bull of the same age, Fred. He'd arrived during a January blizzard as part of a shipment of 11 'heifers.' The previous owner had sold because his land was overgrazed and short of feed. All the animals showed this; they've never gained the stature we breed for. However, the heifers later threw three full-sized offspring when bred to Monster. Gutted, Fred weighed only 910 pounds in the pickup—with head and hide on, compared to 1,000 pounds for Herman with his chipped horn. Maybe Fred would sire calves just as fine as Herman's half-brothers, but he could never look as good as those of his age group sired by Monster and raised together. (Incidentally, none of those smaller cows will go out in live sale; only their large calves.)"

NBA Director, Jerrell A. Shepherd, reports that in his Missouri herd, careful weighing shows that a fraction of each age group weighs more than 10% over the average for the whole age group. His "keepers" for breeding come out of his top group. The rest are subject to the normal herd marketing pattern. This experience suggests that many of us might benefit from a sealing program in addition to trying to estimate sizes by appearance alone.

Careful culling designed to fit each raiser's own circumstances is needed to increase total buffalo numbers and improve the health and vigor of the animals in each herd. Sound culling means hard choices, especially for those of us who raise the animals and know each by name. Continent wide, probably less than 10%, or 7,500 animals, are selected for culling each year—which is why the number of buffalo on the continent continues to increase.

Good culling practices can be summarized as follows: (1) Sickly animals that don't recover quickly should not be kept: the case of Livia, (2) Dangerous animals must be eliminated, a human life is worth more: the case of Cleo, (3) Gradually, selection rules develop so that a fraction of the buffalo leave the ranch each year, helping to support the rest and increase the total supply. Those that survive are improving through sacrifices such as those of Herman and Fred.

9
Insurance
Pointers

by Richard E. Kuhn,
Insurance Agent
owner of Lazy K Farm, WI

One of the necessary evils in life is the need for insurance. When a loss occurs, this evil becomes a need. Your first thought: "Is it covered?" At the time of purchase, many people do not determine their exposure to loss. Farm and ranch insurance is not a mail-order purchase, it's a tailor-made procedure.

This takes time and full understanding. Your independent insurance agent should be given the opportunity to review your entire operation. He can then determine your exposure to major losses and, through use of deductibles and self-insurance of some exposures, save you premiums. Any changes or acquisitions should immediately be communicated to him. Remember—insurance is designed to distribute losses to all participants. Those without losses help pay for those with losses, or the cost would be unaffordable. In the same vein, large deductibles typically help to reduce premiums considerably. Most of us can handle the small "nuisance" claims with a hammer and nail!

Policies need not be a big mystery, buried in fine print. Certain basics are common to all policies. Limits of coverage, what is covered, how it is covered and what is excluded.

The face of the policy will have your name, address, limits of liability and description of coverage, i.e., buildings, personal property, auto, live-stock, workers' compensation, products liability, etc. It will also have the policy period and premium charged.

Attached to this "face sheet" will be endorsements of riders outlining what and how you are covered. Perils are described first, i.e., fire, wind, collision, theft, legal liability, etc.

Next, the exclusions are listed or described. These will vary by the type of policy such as fire, auto, cargo, etc. These are usually followed by supplementary coverages and definitions. Pay attention to the definitions since they often take the mystery out of the fine print.

Most policies covering property, livestock, cargo, machinery and equipment, jewelry and furs, guns, etc., are either on a "named peril" or "all risk" basis. Name peril policies specify the perils that are covered. All risk policies cover every exposure except those that are specifically not covered and listed as "exclusions" in the policy. It is usually better to purchase the all risk policy; to justify the additional cost for this broader coverage, buy a higher deductible. This, in my opinion, is true insurance.

Now that you know how to read a policy, let us review the exposures most common to a farm or ranch. Remember, you still need professional counsel to tailor coverage for your individual needs, exposures, and pocketbook. Believe me when I say if insurance agents could be paid by the hour, as many professional people are, there would be fewer lawyers and more independent insurance agents!

About 25 years ago the insurance industry decided they could do a better job more efficiently if they combined your needs into "package" policies. Instead of a separate policy for each need, they would combine them into one with separate endorsements attached to one face sheet. This actually resulted in a saving to the insurance buyer and improved the coverages available. For example, two of my companies have come out with a FLEXIBLE farm and ranch owners' "package." I emphasize the word flexible because they can do what I mentioned previously, tailor the package to your individual needs. This applies whether you are a small family-owned operation or a large agricultural complex.

The policy can provide protection for the homes and personal possessions, farm and ranch structures and farm personal property. It can shield an insured and family members against a variety of farm and personal liability exposures at home and away. Coverages can also extend to non-farm for ranch business activities.

Built into the policy are many coverages for the main dwelling, farm and personal liability, and medical payments to others. With this feature there is less chance that an important coverage will be overlooked. An insured can also benefit from features such as optional replacement cost coverage for farm buildings and household personal property, and automatic increases in policy limits to match the rate of inflation.

Another valuable feature is the risk variation rating plan. This allows adjustment for rates to reflect those of you who practice good housekeeping and well managed and maintained operations.

Wording of these policies is pretty much in "sidewalk American," easy to understand. Policy contents are organized in a logical sequence and major headings allow fast location of coverages and important information. In reviewing these coverages, pay particular attention to the dollar limits of coverage for certain items. For example, trees, shrubs, and plants up to $500, outdoor antennas up to $200, money, bullion, coins, and bank notes $200; $500 for damage to contents of a freezer or refrigerator, including farm products, due to power failure or electrical breakdown; $5,000 on theft of silverware; $2,000 on theft or unauthorized use of credit cards or bank fund transfer cards, check forgery or counterfeit money (been to a livestock auction lately?); $1,000 on theft of jewelry, watches, gems, and furs.

An often overlooked feature of many of these policies is additional living expense or loss of use of certain rental property. For instance, if you're forced to vacate your home and move into a motel, the cost would be considerably more than living in your own home and making your own meals. This additional cost would be covered under your policy. It will also reimburse you for loss of fair rental value from a covered peril. This is also available in the form of "extra expense" for loss to machinery or equipment, limit optional, in the event the equipment is damaged or destroyed from an insured peril.

Of course equipment, supplies, feed, livestock, and poultry can all be included in the policy with limits that vary, i.e., $2,000 per head of livestock, $2,500 for farm equipment loaned or rented to you. Higher limits per animal or machine can be added by scheduling (describing each item by endorsement). Even private power and light poles are included to $500.

Another form of coverage important to some is cargo insurance. If you transport your own animals or equipment any distance yourself, you have a large exposure to loss while in transit. Let's say you or a member of your family have raised a prize animal and take it to a national show or fair or auction, maybe several hundred miles from home. Enroute the transporting vehicle or trailer is "wiped out" and you suffer a several thousand dollar loss, not counting all the time and effort put into the raising of this animal. For as little as $1.50 to $2.00 per $100 of coverage, with a $100 to $250 deductible, you could be reimbursed for this type of loss.

One area of property insurance often overlooked is loss of earnings or business interruption. This endorsement to your policy is designed to provide coverage for loss of income due to the interruption of your operation due to a peril which you're insured against. For example, if an important building that was an integral part of your operations, is damaged or destroyed, you would probably be reimbursed for the loss to the building. But what about the loss of income as a result of this occurrence? Or what if a large part of your herd were damaged or diseased? What you would have made in income, had this loss not occurred, can be insured against. There are formulas available to determine your potential and you can purchase accordingly. The price for this is surprisingly reasonable. Loss of income can be considerably more than the loss of a building. With continued income you can always rebuild a building or a herd.

Speaking of livestock, there is available a Livestock Mortality Policy or endorsement. This will indemnify you for loss of horses or other livestock, including buffalo, due to their death from natural causes, fire and lightning, accidents, acts of God, acts of men other than the owner or his employees, and necessary destruction. This is actually life insurance for animals! It also includes their destruction as the result of injury or disease. One word of caution. Never do anything but preservative action without having a qualified veterinary surgeon on hand to certify the action taken was necessitated solely by accident, disease or illness. He can certify that every attempt was made to preserve the animal's life.

There are two basic ways to purchase this coverage. One is known as "blanket" insurance on the herd with a maximum per animal. The other is "schedule" basis where you list each animal to be insured and can properly identify by tag, brand number, or vet number.

The premium will vary by type of animal and habitat. In addition to being a general insurance agent, my avocation is raising buffalo on my 650-acre farm here in northeast Wisconsin. I have about 100 head of bison; many are absentee owned. An option I have made available to these absentee owners is insurance on their animals. One of my insurance companies has agreed to a blanket policy with coverage limited to $1,000 per head for an annual premium of $50. I list the ear tag number and the name of the owner.

How about crop insurance? A day may come when weather patterns can be controlled or modified. Until such a time, however, insurance offers primary protection against potential financial ruin because of damage to crops by the elements. The first crop policy was issued more than a century ago with hail as its basic peril. Currently, it is available in all states and is sold by private insurance companies, as well as Federal Crop Insurance Corp., usually through independent insurance agents.

A variety of crop-hail policies are available, each with its own set of coverages and exclusions. While the Crop-Hail Insurance Association has produced a simplified standard policy, coverages generally differ from one state to another. Coverages are tailored to take into account differences in climates and other environmental factors.

Crop-hail policies share common characteristics. In nearly every case, the coverage becomes effective once the crop is above the ground, and expires when the crop is harvested. Also, the limits of coverage are expressed in terms of acreage rather than the total value of the crop. Until recently, private insurers insured only against the perils of fire, lightning, hail and wind (if accompanied by hail which destroys at least five percent of the crop). Now you can purchase "all risk" which, in effect, covers crops against such additional perils as insect, excessive moisture, drought and any other peril not specifically excluded. The passage of the Federal Crop Insurance Act of 1980 made it feasible for private insurers to offer "all risk" policies because they can reinsure them with the Federal Crop Insurance Corp.

Now that we have covered most of the "real" property type of exposures, let us review the General Liability type of losses or Section II as it is called in Ranch and Farm Owners' policies. It is increasingly clear that the maintenance of a farm or ranch and its operation subject a farmer to third party claims for bodily injury and property damage. Two liability policies are available for farming operations.

Farmer's Comprehensive Personal Liability is especially designed to cover the operator of a farm, ranch, plantation, or similar agricultural establishment against all loss from legal liability to others arising out of the personal acts of himself or his family or through the operation of his farm. Animal Collision coverage, Custom Farming coverage, and Employers' Liability coverage (including Employee's Medical Payments coverage) on farm hands, not included in the basic policy coverage, may be added for additional charges.

Comprehensive General Liability is often considered by many agents for the unusually large operations. I do not share this philosophy because of the many and various exposures the farmer has in today's operations. This coverage insures against declared existing liability hazards, and any additional hazards which may occur during the policy term, arising out of the building, premises, and farm operations. Combination farm-cannery operations are an example. Selling meat that has been butchered and processed, then packaged for wholesale or retail sale is another example. Product liability is optional. More on this later. Though lease of premises and elevator or escalator maintenance agreements are covered without special treatment or charge, important contractual agreements such as construction are not automatically covered. The policy must be endorsed to cover them.

If Broad Form Comprehensive General Liability is added to your policy, the basic coverage is extended to include such additional hazards as contractual liability, personal injury (libel, slander, and defamation of character), premises medical payments, host liquor law liability (not applicable in all states), real property fire legal liability (damage to property in your care, custody, and control), broad form property damage exposures, incidental medical malpractice, non-owned watercraft liability, protection for employees as insureds, bodily injury resulting from intentional act, and newly acquired organization exposure. Whew! All of this sounds like insurance "double talk." But when you're served with long winded legal papers asking you to appear in court because somebody is suing you for $1,000,000 you'll be glad the homework was done prior to the claim. You can let the insurance company defend you and pick up the tab.

Even if you're innocent, the average cost of defense today is a whopping $26,000! This is a national average, not New York, or California. Ask any insurance agent or claims adjuster. Picture a happy family cruising your country side when suddenly they plow into your buffalo or other animal and are very seriously injured. Unless you have an "open range law," those dudes will start thinking of $1,000,000 for pain and suffering—loss of consortium—and anything else their attorney can include.

Products liability, alluded to earlier, is usually included in the farm or ranch owners' Section II general liability. But due to considerable variance in products and the methods of distribution, I can't emphasize enough the importance of complete communication with your agent on the nature of your individual operation.

For example, if you do occasional slaughtering on your premises, this would be covered. If it is extensive, you had better have this confirmed in writing as to whether or not this is covered. If you lease part of your premises to a third party to distribute all that good buffalo meat, this is an extensive exposure to a loss and you would most certainly be enjoined in a lawsuit if people should become ill as a result of the food served. On my buffalo farm, since I do not have any cash crop but rely on the income from the meat sold to restaurants in Wisconsin and Illinois, I have a special endorsement confirming the insurance company is aware of this exposure.

Another area of concern is if you allow hunting on your premises. Some big "executive" is willing to pay you a fancy fee to brag about the buffalo he shot! But he may not be the best shot in the world and shoot *somebody,* or the neighbor's prize bull. Same is true regarding pheasant and chukar hunts. I guarantee your name will be on the lawsuit for large numbers!

Workers' Compensation is an entirely different policy and you have to have the expertise of your agent to protect you for your own state. In some, it applies to farm hands and in others they come under your liability policy if they sue. This insures loss due to statutory liability of you, the employer, as a result of personal injury or death suffered by your employees in the course of their employment. The limits of liability are prescribed by your individual state law. If you operate in more than one state, and an employee may live in another state than yours, have an "all states" endorsement attached to your Workers' Compensation policy. When a Compensation Act does not apply, Employers' Liability coverage under a Farmer's Comprehensive Personal Liability is recommended.

Let's talk about auto and truck insurance. This normally involves a separate policy on which all of your vehicles should be scheduled. Again, legal phrases in the policy may seem difficult to understand. They are used because courts have established their meaning through repeated testing and interpretation.

The policy is made up of a jacket or booklet describing the coverages, definitions, and terms used including coverages and exclusions. Then we have the declarations which is a separate page that supplements the policy jacket. It contains your name, address, and a description of the coverages, limits, deductibles, and other items that you select and, the insurance company agrees to provide. Thirdly, we have endorsements which fit special circumstances like state required forms, change of car or truck, etc.

Types of coverage include bodily injury, property damage, medical payments, protection from uninsured motorists, collision, comprehensive, rental, and supplementary payments. Bodily injury liability protects you up to your policy limits against a claim or suit for injury to pedestrians or people in another car, caused by an auto accident in which you are considered legally responsible or liable.

Property damage liability protects you up to your policy limits for damage to property that results from an auto accident where you are considered to be legally responsible. Property refers to another person's car or property like mailboxes, trees, cars, and so on.

Medical payments provides for most medical, surgical, dental, hospital, and certain related expenses resulting from an accident in which you, your family and guests are injured in your car. This also includes coverage for you and your family, should you be injured as a pedestrian. These payments are made regardless of who is at fault.

Uninsured motorists is for "people damage" only. It pays for bodily injury from an uninsured motorist (one who does not carry any auto insurance), hit and run accidents, or for a driver whose insurance company is insolvent.

Collision coverage will reimburse you regardless of who is at fault in an accident. If you are not at fault, your insurance company will subrogate against the other party for the damage they paid you. It will not be charged against you. This is always written with a deductible and I cannot recommend strongly enough that you select as high a deductible as you can afford. Many of my clients, myself included, carry a $250 deductible on collision.

Comprehensive coverage provides payment for damage to your car or truck caused by other than collision, glass breakage, theft, fire, explosion, impact with an animal, plus vandalism. It also is available with an optional deductible; $50 is most commonly used.

When discussing with your agent the limits of liability you should purchase, please listen to his advice. With today's courts being so generous in their awards, I consider $100,000 per person, $300,000 per accident and $50,000 property damage as minimum.

The premium you pay for your auto insurance is based on six basic categories. Where you live: rural rates are usually considerably lower than big city rates. Type of car you drive: I can guarantee a Corvette costs many times more than a Ford Granada or pickup truck. How your car is used: Sales people who travel a lot always pay more. Who drives your car: your inexperienced teenager son or daughter can drive your premium up considerably. Your whole family's driving record can have a large effect on your premium. Many of my clients, be they doctor, lawyer, or Indian chief, have their teenagers pay for the additional cost of adding them to the policy. They learn quickly the value of the privilege of driving! The coverages, limits and deductibles have an effect on what premium you pay. Some companies will add this coverage to your farmowner's or ranchowner's package and include a discount.

Due to the unrealistically high judgments of recent years, the purchase of excess liability or "umbrella" coverage has become very popular. These normally start at one million and I have some in the five and ten million limit numbers. This coverage is just what it means—excess over all your other underlying policies. Normal retention or deductible is $250 for personal and $10,000 for the larger commercial operations. Prices are reasonable depending on the limits on your underlying coverages and the type of operation you run.

Accident and health insurance is something I am sure all of you carry in one form or other. With the astronomical cost of any type of medical treatment today, nobody can afford to be without this coverage. It provides twenty-four-hour coverage as opposed to worker's compensation which covers only work-related injuries. If you have three or more employees, I suggest you obtain several quotes. But be careful, the cost of this coverage is often predicated on the type of coverage purchased.

For individuals and single family clients I sell a Major Medical form from one of my companies. Very briefly, it pays on an 80/20% basis up to $3,000 and then pays 100% after that up to $1,000,000. This means the out of pocket costs to you would be $600 (20% of $3,000) plus your deductible. Again, take as high a deductible as you think you can afford. It really reduces the cost and the savings to you can be used when the time comes to pay the deductible. The deductibles available range from $100 to $1,000 per calendar year. One caution on accident and health policies, both individual and group: your children are covered only to age 23 and then only if they are students. There are exceptions, so have your agent advise you. Also, pay the premium quarterly and save money.

Disability income insurance is an often overlooked coverage and in my experience has saved home and family for many of my clients. It will guarantee you an income in the event of injury preventing you from making a livelihood and is often combined with Social Security benefits. Since the definition of "disability" is the key to this policy, review this paragraph in the policy very carefully. I've seen some where, if you are well enough to sew buttons on shirts, the company considered this an income.

The premium is based on your age, occupation, waiting period, and how long you will continue to receive payments after injury. Waiting periods can vary from seven days to six months. After the waiting period, you can receive weekly payments for various periods of time while recovering from sickness or accident. The most common is two years for accident and sickness, up to five years for an accident, and lifetime income from sickness.

Have you and your spouse made a will? I never cease to be amazed at the number of highly paid professional people who have not taken the time to perform this simple task. All I ask you to do is have a tax accountant show you what would be left of your hard earned estate after taxes should you die "intestate" (without a will). It will cost you nothing to have your insurance agent do an estate plan for you and show you how to shelter your assets from taxes. Current tax laws are lightening the burden for farm and ranch owners, but professional counsel is still a necessity. Do it now!

This brings up the subject of appraisals. I'm sure if you're like me, when you go to the bank they want a financial statement prior to lending you any money. This is one form of appraisal of your assets. However, what is good for the bank is not necessarily good for Uncle Sam's tax army any more than the value you place on your real property (building and contents and equipment). With professional advice prior to your death, you can leave considerably more to your survivors. If only the good Lord would let us know in advance when we are going to die, it would make estate planning very simple!

Appraisals for insurance are another specialty. They do not have any relationship to the market value of your farm or ranch or land. What it would cost to replace your buildings and equipment today—minus depreciation—is known as insurable value. Eighty-five percent of the people today are underinsured for fire, wind, hail, and so on. Ask yourself what it would cost to replace your barn today and then check the amount of coverage you carry on your policy. In the event of loss, guess who is going to pay the difference?

Solar Energy equipment is not something we all have on our farms at present but as improvements in this field progress and the price comes closer to our pocketbooks you should be aware that coverage is available. Basically, the coverage provides direct physical loss or damage to all solar energy systems described in the policy. This would include such items as collector units and conductor panels; heat transfer and exchange mechanisms; plumbing, piping and duct work; circulating medium; control and safety devices; storage units; and other related equipment. Exclusions are pairs and sets or items with several parts where the insurance company will pay only for the lost or damaged part. Also excluded is wear and tear; gradual deterioration; discoloration, deterioration or corrosion of solar absorption panels; corrosion or rust; and dampness, dryness, cold or heat.

One caution in purchasing this coverage is to be sure in purchasing an amount equal to 100% of the actual cash value. If you, for example, only insure to 50% of the total cost of your installation you could be greatly penalized in the event of even a partial loss. Values should not be difficult to determine since your installation should be fairly current.

Personal mini-computer or data processing equipment is also now available in a special tailor-made policy. It includes equipment you own, rent or for which you are legally responsible. Data processing equipment means a network of machine components capable of accepting information, processing it according to plan, and producing the desired results. It includes hardware. It does not include software.

Crime coverage is not something every ranch or farm operation needs since most of the money handling and accounting is done by some member of the family. However, if you are diversified or have a feed mill or warehouse operation in conjunction with your farming involving several employees, you should consider a comprehensive dishonesty, destruction, and disappearance policy. This provides all-risk protection for money and securities on and off the premises, caused by dishonesty, mysterious disappearance or destruction. It provides insurance against loss due to dishonesty of employees, loss of money and securities within and without the premises, damage done to premises and equipment, loss of securities in safe deposit or through forgery of outgoing instruments, safe burglary and messenger robbery on other property. It is often wise to have an employee with fiduciary capacity bonded, for psychological reasons alone.

Safety and good housekeeping were not left until the end because they were not important. Most good farmers and ranchers are well aware of the hazards of their operation. However, there must be many who do not practice this or the rate for workers' compensation would not be as high as $16 per $100 of payroll. Neither would there be the high fire and allied perils rates for property insurance if more would make good housekeeping a habit. Wiring, combustible liquid use and storage, equipment maintenance, careful observation of new hired hands while performing their duties all require constant vigilance. It can also go a long way towards reducing your overall insurance costs.

I highly recommend you purchase *Farm Accident Rescue,* catalog number NRAES-10 from Riley Robb Hall, Cornell University, Ithaca, NY 14853. It could save a life!

As I have emphasized several times, you can save many premium dollars by seeking professional advice no matter how small your operation. If I can be of any assistance, please feel free to communicate with me. I would like to credit the following companies for their assistance in preparing this chapter: St. Paul Fire & Marine Insurance Co., Fireman's Fund-American and the Hartford Insurance Group.

10
Registering
Buffalo
= Increased Profits

Breeders of purebred cattle, horses, etc., have done themselves and their breeds a lot of good by establishing registries. Some breed association herdbooks go back hundreds of years, giving the buyer of a registered animal documentation that the animal's genes are untainted and that it carries the bloodlines of outstanding individuals. That animal commands a bigger price, and will get more valuable progeny, than one of uncertain ancestry lacking distinguished forebears. Registration of domestic stock has been a powerful tool in improving the breeds. Registry will do the same for *B. bison*.

A couple of attempts have been made to establish a buffalo registry, but the criteria are deemed by some as too lax. NBA is working with other groups to establish a truly meaningful bison registry that will do for this wild animal and its breeders what registry has done for domestic breeds.

There are several improvements and refinements being considered for the National Buffalo Association registration program. A point system, based on the ABA judging criteria, is also being considered to establish minimum quality standards. The program we now have is a good start in providing lineage and pedigree information for those who want to make use of it. The reliability and integrity of the registration program ultimately falls back upon the honesty and cooperation of the individual breeders involved.

Following are two informed views on establishing a buffalo registry.

NBA HELPING TO ORGANIZE NORTH AMERICAN BISON REGISTRY

by Mike Fogel
Fountain City, WI
NBA member and buffalo breeder

In a registration program, we have to first decide what we hope to accomplish. For the benefit of buffalo raisers who have never used a pedigree or produced purebred, registered animals, registration is a valuable tool for improving the quality of your animals and, after a time, seeing the results.

NO LONGER ENDANGERED–THANKS TO RANCHERS

We think that just to propagate the species and not improve it is a mistake. Now that ranchers have rescued the American buffalo from the crumbly edge of extinction, where it teetered in 1900, and brought it back to an excess of 75,000 individuals in the U.S. alone—we think it is time to start making it better. A registration program and established bloodlines, as in other breeds, can open new markets and enable the small breeder to use his animals to better advantage and receive prices in accordance with his efforts. It would be advantageous to the serious breeder to be able to exchange established bloodlines. We could get into genetics and linebreeding.

First, let me briefly outline the North American Bison Record:

A number of people have now divided their herds into small breeding units permitting the use of one known bull on a designated bunch of cows. This is known as "single sire mating." (If more than one bull runs with the cows, you never know which bull sired which calf.) Single sire mating is necessary so that the paternity of any given calf will be known for certain.

While this practice would not be practical for entire large herds, it is not necessary to maintain registry on the entire herd. Larger ranchers would simply cut out the top 10% or so of their animals for division into single sire bunches. Known parentage can give the rancher a premium market for his top animals.

Some people don't agree with the philosophy of one bull in a cow herd, but for a worthwhile registration or breeding program, it is vital to know exact ancestry. Parentage can be proved only with blood typing. In our program, we not only want to know our animals are pure but also their exact lineage. When we choose a bull to build a foundation on, we want to make sure the calves carry his genes, not some inferior bull's.

This registration program is an optional service so progressive breeders can improve the breed. It could also open the doors to purebred sales, shows, and marketing service based on a herdbook such as the recognized breeds maintain.

BASED ON SCIENTIFIC BLOOD TYPING AND RECORDS

The science of blood typing buffalo is fairly new, but the procedure does establish parentage beyond doubt. Two blood typing laboratories have agreed to work with us: Texas A & M and Stormont Laboratories, Inc., in California.

The buffalo industry is growing and careful records and pedigrees will have to be used if we want to compete for our share of the market. As in cattle, or any other registered breed, the offspring of the proved top producers command the best prices. With a registration program, blood typing and lineage records, we are eliminating some risk for ourselves and for potential buffalo raisers. If you can show buyers that your blood-line is consistently producing more meat, faster gain or better heads, you've got an edge over someone who has no records.

REGISTER JUST THE BEST

As indicated above, there is no need to register all buffalo any more than there is need to register all Herefords or all Quarter Horses. Just the best should be registered. This means we should register only our buffalo that have the traits we would like to pass on to the offspring. Dr. Stormont tells us that, through blood typing, he would be able to help choose the bull that would have the best range of breeding possibilities, that would best 'nick' with the outstanding traits of a given cow, by comparing blood samples. Blood typing will also help keep people honest. Without complete accuracy, any registration program is worthless.

Another point to consider is: Who is to say what a good buffalo is? Your idea may not be the same as mine. One breeder favors a long, rangy buffalo with a dark coat. Another may want a short, compact beast with a big hump and light wool. We don't have a standard set yet as do the Angus and Hampshire breeders. Now's the time for the industry to develop a standard.

BLOODTYPING TO DETERMINE PURITY AND PARENTAGE

by Clyde J. Stormont, Ph.D.
Stormont Laboratories, Inc.
Woodland, CA

It is calculated that trillions of blood types are possible in cattle, but only about 40,000 in bison. Two blood typing tests are available through Stormont Laboratories to determine your buffalo's purity and parentage. As the NBA plans for a Registry come closer to materializing, blood typing becomes a more important tool for the bison rancher. Here are the two Stormont packages:

Package 1

This combination of electrophoretic and serologic tests ascertains that an animal being certified to enter the bison herdbook or registry is in fact a purebred bison. It detects any taint of genetic markers that could have been derived only from cross-breeding with domestic cattle in some

Dr. Clyde Stormont.

previous generation.

The object of this package is to detect and weed out from the Certified purebred bison any animals that have as little as 1/64 to 1/512 domestic cattle blood.

Package 2

This is the ideal and most economical set of tests for any animal whose stated parents already have their blood types on record. It includes tests for all segregating blood factors plus the 6 CA types. It is designed primarily to detect errors in parentage assignments and to rectify those errors.

COLLECTING AND SHIPPING THE BLOOD SAMPLES

Your veterinarian will usually have a supply of the lavender (purple) stoppered Vacutainer tubes which draw up to 7 ml of blood. If he has only the Vacutainer tubes which draw up to 3 ml, then ask him to take two tubes from each animal. These lavender stoppered tubes contain EDTA as the anticoagulant.

Prior to collecting a blood sample, it is most important to identify each animal by name, registration number—if registered—or by tattoo, ear-tag or brand. Enter this information on the label and proceed to collect a blood sample either from the jugular or tail vein. Also enter this information on the Blood Typing Record of Identification forms to be mailed along with the blood samples. These forms will indicate whether a blood sample is being submitted for certification or parentage verification.

Chill the samples in the refrigerator (DO NOT FREEZE) for about two hours before packing. Ship in a styrofoam container designed for these tubes, or wrap in paper and pack into a strong carton. Ship by *Priority Mail* to Stormont Labs, 1237 E. Beamer St., Suite D, Woodland, CA 95695.

WHY REGISTER BUFFALO, ANYWAY?

A registry must do more than establish that the critters are buffalo. For the most part, you can look at 'em and tell that.

LOCKING-OUT THE FREELOADERS

A registry must have integrity. For example, a person must not be able to buy a cow somewhere, hurry home and register her by establishing her as a buffalo, and then turn around and make a big deal about this being registered stock and therefore more valuable than grade stock.

What is the object of registering buffalo? Is it to give the owner a fancy set of papers so he can resell that animal at a fancy price to an uninformed buyer? Or is it to upgrade the breed and guarantee a purchaser he is buying superior stock? The NABR assures the latter.

"GRANDFATHERING-IN"

Any problem with buffalo previously registered under inadequate systems can be easily solved by "grandfathering" them into a new program. It is doubtful much damage would be done by grandfathering these few in. On the other hand, it would be a gross error to continue along inadequate lines just because that was the way it was done before. With the exception of the few registered animals on record, we're starting from scratch. Repeating mistakes is not going to give any registry a fresh start.

MUST SET STANDARDS

We must establish the purpose of registration. Registration must be more than just certification that the animal is a buffalo. A registered buffalo should be something special, so that 10 years down the road, that herd will have much added value. If it is too easy to register, the registry is not going to be worth the paper.

CERTIFICATION FIRST

We must establish what is a superior buffalo. Once a criterion is set on the traits registered buffalo will perpetuate, then you can set up a registration program. Certification comes first. That is essential to determine you're dealing with pure buffalo. Then you pick the best of what you have for breeding stock. Certified buffalo bred to certified buffalo get certified calves. These calves grow up and breed with other certified stock; their progeny carry the traits qualifying for registration.

But just because a calf comes from certified stock, it shouldn't automatically be eligible for registration. It's got to show the desired traits. Which means someone has to inspect and approve it.

ONE SIRE HERD ESSENTIAL

Keeping a one sire herd is essential. Keeping accurate records of what calf came out of what pair is essential. This should be carried on for several generations to assure the traits wanted are stable in this population. The sire and dam records, along with certification records of base stock going back the required number of generations, should be provided with application to register. Then you have an established pedigree of sorts.

When the conditions for registration are met, you have something of value. You have registered breeding stock. Buyers can either start from scratch with a certified herd and wait, as you had to, or they can save a lot of time by buying registered stock of much higher value and price. If registration can be obtained swiftly by any applicant, there is no value to it. If registration is worth having, it must be worth working and waiting for. In the meantime, the animals that have been grandfathered in will have a decided advantage over the certified stock, but they won't have a pedigree and if the buyer is informed, they won't have much more value than certified stock. Here is where industry members must police themselves.

NBA will accept grandfathering in with a cutoff date of September 30, 1982, to prevent people from hurrying up and registering a herd so they can get immediate status and attain an edge over the person willing to work hard for something special. From this point forward, everyone must be on equal footing. The NABR assures this.

NORTH AMERICAN BISON REGISTRY
CERTIFICATION AND REGISTRATION PROGRAM
JUNE, 1983

PURPOSE

The purpose is to better serve the needs and demands of buffalo breeders and the buffalo industry, as well as help insure the purity of the breed and the genetic propagation of superior seed stock for continued upgrading of buffalo breeding programs, to authenticate the purity and parentage of buffalo so that superior bloodlines can be identified and recognized, to lend credibility to a selective breeding program, and to provide a two-step approach which gives buffalo breeders an option to fit the needs of their individual programs.

GENERAL PROCEDURES

1. Breeders are encouraged to select only their top animals for regis-

tration as these are the future seed stock producers. Animals found to be unworthy of registration are subject to review by the Registration Committee of NABR with the possibility of revocation of papers if they do not meet minimum size for age and quality standards.

2. All animals eligible for registration must be assigned an individual private herd number and name that is unique only unto them. Breeders must submit their private herd codes to the NABR office (i.e., HBR, BJR, BX, etc.). They will be approved on a first-come, first-served basis. There will be no duplicates so breeders should submit a first and second choice.

3. Animals must carry their identification in the form of a tag or tattoo so that they can be positively identified. Identification such as HBR Female will not be accepted; however, HBR Female 82-10 will be, or HBR Rosie 82-10.

4. The first option will be called certification. This procedure will replace the former ABA registration program and will follow basically the same rules and regulations that applied to that program. All animals currently registered will be issued new registration papers using the same number already assigned to them. Cut-off date will be September 1, 1982. After that date, all animals submitted must qualify under the new programs.

5. The certification program will serve as the means to document the purity of the animal through a blood typing procedure whereby animals will be certified 100% bison bison. Offspring of certified animals will not have to be blood typed to qualify for certification.

6. The registration program will be the second option and will require, first, that all animals are certified, and second, that positive sire and dam identification and accurate birth date information be provided. Single-sire mating of bulls with dams will be mandatory to positively identify sires. At this time, animals produced from multiple sire breeding will not be accepted regardless if all sires and dams are certified.

CERTIFICATION PROGRAM

1. Owners must request certification applications from the NABR office. Upon completion of the applications, the white copy should be sent to the NABR, along with a check for the certification fee.

2. Upon receipt of the money, the NABR office will issue blood sample vials to the owner, whereupon he/she will take the blood samples and send them, along with the pink copy, to Stormont Laboratories, Woodland, California.

3. After the blood analysis work is completed, the results will be sent to the NABR office whereby a certification number will be assigned and a tag issued for the animal. Certification papers and tags will then be sent to the owner.

4. The owner must then put the tags on the animals as proof of their purity as 100% bison bison.

5. A local licensed veterinarian or a designee of the NABR Registration Committee must be present when blood samples are taken and the data recorded to insure that proper anti-contamination procedures are followed. The expenses and fees of this person shall be paid by the owner.

6. Owners may place the NABR certification tag in either ear of the animal but should take care in application so that the tag will not be torn out or lost.

7. The data forms must be signed by the veterinarian, or NABR designee and the owner. It is the responsibility of the owner to properly care for, package, and send the blood samples to the appropriate laboratory.

8. If the blood analysis reveals an animal to be other than 100% buffalo, certification of that animal and its subsequent offspring will be denied.

9. When an owner disposes of a certified animal, the NABR office must be notified so the file can be updated.

10. If a certified animal is sold, the owner, at the time of the sale, must send the certification paper to the NABR office, along with the transfer fee, and indicate on the back of the paper the name, address, and telephone number of the new owner. Record of the ownership change will be made by the NABR and approved by the signature of the executive secretary or the registrar. The paper will then be sent directly to the new owner. Producers should note that it is the responsibility of the seller to pay for and activate the transfer of sold animals.

11. If a buffalo is killed or dies, the owner must notify the NABR office whereupon the paper will be placed in a dead file.

12. Offspring from a certified sire and dam may be certified without a blood test. They may also be eligible for registration, providing positive proof of sire and dam is available.

13. NABR reserves the right to demand a blood test, at the owner's expense, on any animal that may appear questionable as to purity. If the said buffalo proves to be impure, the certification will be dropped and other actions may be taken if fraud or intent to defraud is apparent.

REGISTRATION PROGRAM

1. No animal will be eligible for registration unless both parents are certified.

2. Only animals whose parentage can be positively identified through single sire mating will be eligible for registration. Offspring from multiple-sire breeding will not be eligible. In the future, the NABR may consider blood typing for sire identification, providing that owners are involved in a total herd blood typing program with a professional blood analysis laboratory, and providing the laboratory feels their accuracy in predicting the exact sire is beyond question.

3. Owners should request registration forms from the NABR office. Forms should be completed and returned to the NABR office, along with registration fees. Upon receipt of registration fees and forms, tags and registration papers will be issued.

4. Registration tags should then be placed in the right ear of the animal.

5. If animals are sold, the owner is responsible for sending the papers, along with the transfer fees to the NABR office, who will in turn record the new owner, authorize the transfer and forward the papers to the new owner.

6. If animals are killed or die, the NABR office must be informed so that the information can be noted on the animal's record.

7. Owners must provide accurate birth dates and should provide birth weight information on each registered animal so that more accurate records can be established.

8. Application for registration must be made only on the recognized forms issued by the NABR.

9. All animals presented for registration must be positively identified by means of an ear tag or tattoo. A herd code must also be indicated and animals given an individual herd name or number as identification.

10. Animals admitted for registration, after meeting verified parentage and certification requirements, must also be sound and worthy of registration as breeding stock quality.

11. Any animal found to be of inferior quality by a designated committee of qualified buffalo producers shall be subject to revocation of registration papers and also those issued to subsequent offspring.

12. The NABR reserves the right to demand a blood test, at the owner's expense, on any animals if questionable parentage is indicated. If the animal is found to be legitimate, the NABR or person bringing action shall reimburse the owner for blood typing fees plus an additional $25 handling fee per animal. If, however, the animal is found to be ineligible for registration, owners will pay a $50 fine and be placed on probation for one year with all rights to enter association sponsored sales suspended. Public notice of the infraction will also be made.

NBA PARTICIPATION

1. The NBA agrees to co-sanction the NABR in cooperation with the ABA. The NBA further agrees that the NABR will be administered by the ABA office under the direction of the ABA executive secretary and the NABR Registration Committee chairman. All fees will be collected and retained by the ABA and all liabilities of the NABR will be the responsibility of the ABA.

2. As co-sanctioner of the NABR, the NBA will be provided with a duplicate set of certification and registration records for every animal certified or registered under the program, within a reasonable amount of time.

3. The NBA will be notified in advance of any changes recommended in the NABR and future co-sanctioning of the program will be dependent upon the approval of the board of directors and membership of the NBA concerning those recommended and/or approved changes by the ABA.

4. The NBA agrees to continue its cooperative effort with the ABA and will not institute an independent registration program unless all cooperative efforts fail or the preceding rules, regulations and requirements are not being carried out in good faith.

5. Finally, the NBA agrees to work toward a registry program which meets the needs of the entire North American continent, specifically the membership of the Canadian Buffalo Association, and will refer all interested parties to the NABR office.

Your
Animal's
Health

III
III
III
III
III
III
III
III
III
III
III
III

11
Helping
Your Buffalo
Stay Healthy

Buffalo are basically robust, healthy critters. They survived everything short of tornadoes and the buffalo killers. But the blessings of civilization restrict their free choice of diet, crowd them into recycling their own parasites, and introduce new hazards from domestic livestock. These artificial conditions demand that you offset their effects.

It all boils down to nature. She superbly endowed these great beasts with instincts and hardihood to fend for themselves if we'll just let them do their natural thing in a natural environment.

This reminds us of the wisdom of Kiowa Chief Kicking Bird, who asked 150 years ago, "Why kill off the buffalo—that take care of themselves—to make room for cattle that *you* have to take care of?"

Roy Houck observes, "We older breeders are probably to blame for the misconception many newcomers to buffalo have. We've played up buffalo longevity and resistance to disease—which are true—but people don't understand buffalo are wild animals, so they put them in confinement and feed them a foreign diet where they can't select what they need. And in confinement, they recycle their parasites. Under those conditions they become subject to diseases and parasites."

CLOSTRIDIUM CAN DO YOU IN
by Bill Moore

We vaccinate all our animals and our absentee owner's animals with "7 Way." Our prime reason for using 7 Way is to protect against clostridium. It vaccinates for four forms of clostridium. Overeating disease, blackleg and lockjaw are all forms of clostridium.

On our ranch in northwest Wyoming, our problem has been clostridium-D, technically called perfringens. We've had buffalo succumb to this and found it very hard to diagnose at first. Once we knew what was happening, how and what to look for and how to prevent it, we cut our losses significantly.

119

As has been pointed out, buffalo are more similar to sheep than cattle. Clostridium is basically a sheep disease. But because buffalo are generally looked at—especially by veterinarians—as bovines similar to cattle, very often the original diagnosis will be "bloat." Unless a post-mortem is done immediately upon the death of the animals, you can plan on them calling it bloat. We know that buffalo don't succumb to bloat, but the vets don't know this.

Animals that die from clostridium generally go swiftly. They are usually your prime, showiest animals, which compounds your loss. You'll see them lying down . . . and within an hour they'll be dead. It's that quick. The disease, or virus, is inherent in the animal. Symptoms and death are triggered by exceptionally lush pastures, which we have at the Bar-X. The reason death is usually attributed to bloat is that, immediately upon dying from clostridium, the gas buildup that caused death makes the blood vessels burst under stress, causing a swelling that looks like bloat. The only way to know for sure what has happened is to have that animal posted immediately.

If an animal dies of clostridium, don't plan on salvaging the meat. It very swiftly smells so bad you won't want to feed it to your dog. We get them out and haul them to the dump as soon as we can. There is no salvage. You can prevent this tragedy with vaccination. Once your vet knows what your problem is, he can help you prevent it. There is no protection except prevention.

12
The
Brucellosis
Controversy

Many cattle people, as well as buffalo breeders, are challenging the brucellosis test. The test is not for the disease itself but for the animal's resistance to it—for the antibodies against the disease. So, if an animal has thrown off the disease—or even has been vaccinated—the test shows he's a "reactor" because he has high *resistance* to brucellosis. But because he's a reactor, he must be destroyed! The test is not for the disease at all, it's for the antibody titre—for the degree of resistance.

The federal brucellosis eradication program is coming under increasing fire from cattlemen as well as from buffalo raisers, who wonder if results offset the side effects. Some cattle ranchers are understandably nervous about the possibility of buffalo transmitting brucellosis to their cattle. NBA has for years offered a standing reward of $1,000 to anyone who proves that buffalo ever infect cattle with brucellosis. So far, no claimants. And scientists have discovered that wildlife forms ranging from earthworms to elk carry the organism. This chapter assembles a wide range of fact and opinion on brucellosis.

"We've gone back to calfhood vaccination for brucellosis on our ranch. Not that we think we need to, but it gives us freer movement in interstate shipment. *Don't* vaccinate bulls! It's not necessary and sometimes it makes them sterile," says Roy Houck. "Bulls don't get it enough to do any damage. They can't transmit it.

"Cows get brucellosis only by mouth. Buffalo get it by licking a dead fetus or discharge that's infected. Calves can get antibodies from the placenta or colostrum (first milk). The calves are born not with the disease, but with resistance to the disease. The main thing," recommends Houck, "is to keep this inherited resistance alive and strong in them."

FEDERAL BRUCELLOSIS PROGRAM IN STATE OF FLUX
Report from Judi Hebbring
NBA Executive Director

We're not exactly sure what is in store for the future, but the following will chronicle the events that have transpired since USAHA met in St. Louis in October, 1981.

The October USAHA meeting, specifically the Brucellosis Committee meeting, was one I was supposed to attend as a member of the Brucellosis Industry Advisory Board. I did not attend, as I was never notified of the dates nor the meeting location. This lack of notification led me to believe my presence was not welcome. I assumed the meeting was "business as usual" with APHIS controlling the meeting and getting their way on most issues. I have not spoken with anyone in attendance at the meeting so I can not report on decisions made or policies developed at that time. I can only report on events that have transpired since—and they have been many.

We are in possession of a copy of a letter sent to Dr. Paul Becton, Director of the National Brucellosis Eradication Program from the state veterinarians of the five states in the Old West Region: North Dakota, Montana, Nebraska, South Dakota and Wyoming. These doctors of veterinary medicine, charged with overseeing the health of the livestock in their states, say in part: "This statement is in reference to USDA/APHIS Veterinary Services request for comments on the proposed changes of the Brucellosis Uniform Methods and Rules, dated Oct. 26, 1981.

"Since this draft is substantially different in content from the USAHA adopted Brucellosis Committee Report, Oct. 16, 1981, the states of the Old West Region aforementioned, met on November 23, 1981, to discuss the ramifications of the proposed changes should they be adopted by APHIS. After a thorough discussion of the specific sections, the five states, as a region, have drafted comments stating their concerns on the proposed changes.

"It should be noted that there is total agreement in the regional response to all sections. Moreover, the region is outraged with this draft for many reasons." The letter then goes on in length to pick the draft apart, bit by bit. Some major points of disagreement are: rigid and inflexible requirements that livestock producers will reject. The fact that inflexibility has been one of the major points of disagreement over the 40+ year history of the program makes the new rules a step backward, not forward. The standards to attain and maintain new state classifications are extremely difficult for states to meet. Increased record keeping burdens on individual states to maintain status and the expenses that will be incurred as a result.

Efforts of APHIS staff members to sway opinion at various committee and subcommittee meetings being evident and APHIS staff members railroading unpalatable points to the committee as a whole were objected to. "At St. Louis, members of the region met to discuss the ill-conceived 16 point recommendation and are prepared to offer strong resistance to the incorporation of most of these points in the Uniform Methods and Rules. The region finds it extremely objectionable to incorporate these compliance recommendations under the guise of "epidemiology." The region is dismayed with APHIS' dogged reliance on the use of the brucellosis card test without use of supplemental serologic tests in vaccinated cattle; . . . results in overcondemnation of vaccinated cattle; . . . erodes industry's confidence in vaccination; . . . penalizes producers who continue to protect against the introduction of brucellosis in their herds; . . . overcondemnation is occurring in areas with the lowest levels of infection in the United States; . . . will result in reluctance to vaccinate because of fears of persistent titers.

"States will fight any attempt to downgrade or complicate vaccine usage. Regional epidemiologists should have more authority to make decisions on herd histories; . . . present document limits that role; . . . too ambiguous; . . . Regions' opinion that Oct. 26th draft does not offer many incentives for a state to achieve "free" status; . . . provisions penalize herd owners rather than reward them. It is questionable to the region whether the advantages of free status offset the restrictions and penalties imposed by these provisions. . . . The region demands are not reflected in APHIS' final and published form of the UM&R." And this is from an area that has little brucellosis infection to contend with.

Imagine the reaction from hard-pressed states! Imagine how hard these provisions would be to live up to for buffalo producers! The letter goes on for pages and pages to pointedly argue provisions of the draft proposal scheduled to go into effect on January 1, 1982. We suspect this letter, and perhaps others of the same nature, was responsible for a press release issued by APHIS in early January. It was titled "USDA Temporarily Postpones New State Ratings for Brucellosis." In that press release, Dr. Becton was quoted as saying, "more time is needed to evaluate data received by various states to determine status under the new system." How's that for an understatement?

On the battlefield in Texas, the state that will suffer the most with the implementation of these new rules, opposition is fierce. Even before the new rules were dreamed of, there was little support of the program. The feeling in Texas is this is a punitive program. If test and slaughter really works, why does the 1982 program demand three tests? If the current vaccine is effective, shouldn't one test be enough? This inconsistency was brought to light recently when Dr. Becton, Director of the Brucellosis Eradication Program (Animal & Plant Health Inspection Service—APHIS), testified before the National Cattlemen's Association subcommittee on brucellosis that none of the tests in use were 100% accurate. When asked how can you prove your cattle don't have the disease, Dr. Becton replied, "Test them." If the test is inaccurate, what does that accomplish? Dr. Becton also testified at this meeting that even if brucellosis is eliminated in an area, it won't be eradicated. Texas cattlemen are looking closely at these inconsistencies and contend that the program is unconstitutional.

Burton Eller, Vice President of Government Affairs for NCA, recently stated, "The old American National Cattlemen's Association was a prime mover responsible for the formation of the Brucellosis Technical Commission in the mid 1970's. It had become apparent the Brucellosis Eradication Program was eradicating more *cattlemen* than Brucella organisms!

"Since the BTC report in 1978, the NCA has worked long and hard to develop a brucellosis program designed to control, and then in the future, hopefully, eradicate brucellosis. As you know, there's a world of difference between the two approaches.

"The current program recognizes, finally, the value of continued vaccination. NCA has never wavered from such a concept, although USDA had forced us away from vaccination by the late 1960's using cooperative funds to the states as a lever."

I find it interesting that APHIS officials cite conclusions of the Brucellosis Technical Commission whenever they need to justify the eradication program. Could it be that they are not leveling with us or are taking statements out of context? If the BTC's conclusions were for a control program, why don't we see one today? Are they afraid that with effective control, which is within reach, the eradication program will be no longer viable and they'll be out of a job?

In 1934 a dairy state U.S. Senator created the Brucellosis Program with the statement, "There are approximately 3,000,000 cows in the U.S. suffering from contagious abortion (brucellosis) . . . either this program is justifiable on the ground of public health or it is not. It is justifiable on the ground of public health and it follows that the expense should be borne by the (U.S.) Treasury." The public health issue that Senator referred to was undulant fever, a disease that was dramatically controlled in this country by pasteurization. Yet, the U.S. Congress recently appropriated over $92 million for the Federal Brucellosis Eradication Program for fiscal year 1982. Why?

The public health threat is long gone, controlled by different means. After 47 years of the Federal Program, those 3,000,000 infected cows are still out there. What really has been accomplished in all these years with all those millions of dollars? Many are asking themselves this question. Governor Clements of Texas is among them and has come out strongly against the Federal Eradication Program.

In New Mexico, a judge recently ruled that the brucellosis regulations promulgated by the Livestock Board of that state were vague and indefinite and did not sufficiently apprise New Mexico ranchers of what the regulations required of them. The judge advised the board to "get their house in order" and recommended that they secure legal counsel to assist them in conforming their regulations to federal constitutional law.

The controversy involved the authority of the New Mexico Livestock Board to restrict shipment of cattle in interstate commerce until certain brucellosis tests are made.

In effect the judge determined: 1) The federal government (Paul Becton and aides) had no business in New Mexico. 2) It was a states' rights issue. 3) The New Mexico Livestock Board had no specific rules and regulations to go by. 4) The ranchers had no specific rules and regulations to violate. 5) If the New Mexico Livestock Board wants brucellosis regulations they will have to make them. And 6) All brucellosis tests are unreliable so the judge is to review all tests before he will resort to condemnation of cows as being diseased. Shortly after this ruling, the entire New Mexico Livestock Board was fired. Draw your own conclusions.

Testifying in this case for the government was: Dr. Paul Becton; Dr. Robert Pyle, New Mexico State Veterinarian; Dwayne Massey, New Mexico State Inspector; and Dr. Cohn, Federal Veterinary Officer. Dr. Paul Nicoletti, of the University of Florida, a brucellosis research person of long standing, testified that the eradication of brucellosis was absolutely impossible. Jay Coyle of Doane Western, a large farm and management service, testified that the cost of ranchers of the federal government's "Uniform Methods & Rules" eradication program would bankrupt and eradicate the ranchers if adopted and enforced.

An item of news that specifically addresses the program as it applies to buffalo is this excerpt from a letter signed by John R. Block, Secretary, USDA. "The U.S. Department of Agriculture (USDA) recognizes the inconsistent treatment of bison calves versus cattle calves in the regulations addressing restrictions on interstate movement of animals because of brucellosis. The Animal and Plant Health Inspection Service (APHIS) is currently reviewing the regulations and is considering proposing changes in the area of your concern."

Now, this would be very good news if it were "new" news. The truth is, those inconsistencies were pointed out to Dr. Becton when he addressed our NBA meeting in Houston in 1979. At that time, Dr. Becton told us this was an error, an oversight in transcribing and the provisions for buffalo should be no harsher than they are for cattle. He said these inconsistencies would be rectified in short order and, since there was no tightening of regulation, he felt it could be expedited in a few months.

Shortly thereafter, the NBA office inquired as to the progress of this amending of regulations. We received no reply. We asked congressional representatives to inquire on our behalf and they were informed that the problem was recognized and measures were being taken to solve the problem. Two years later when I met with Dr. Becton in his office in Maryland, I again asked what the status of this inconsistency was and was informed that, with a new administration in charge, new criteria were needed to effect changes. This sounded like double talk to me, but who was I to argue?

Now three years later, with continued congressional pressure, we've finally got someone to admit there is an inconsistency in the regulations. With any luck, this inconsistency will someday be rectified. But in the meantime, how many producers are losing valuable income from the loss of sale of calves? How many dollars are involved here because someone didn't transcribe the original regulations properly? If we had been given the opportunity to comment before this regulation became law, it probably wouldn't be law now, but this wasn't sent to buffalo producers for comment: it was sent to cattlemen and state enforcers.

This is one of the basic points of our contention. Buffalo producers have never been asked what regulations they can live with. The regulations have been arbitrarily imposed upon them and then they are expected to live with them. When they can't, it's next to impossible to effect a change once it's recorded in the Code of Federal Regulations.

When a producer does squawk, APHIS cites two other producers who are in compliance and will testify to the validity of the program. This carries a lot of weight with congressmen and senators. The problem is, not enough people are willing to squawk. Most assume that the law is the law and there isn't much you can do about it, except to learn to live with it. That's not true. But you must let your representatives know how you feel before any changes will occur.

The whole subject could be moot with new developments occurring. Recent headlines state: "President saves millions by shaving $$ from Brucellosis Program." The articles goes on to say that the goal was set to eradicate this disease but it will cost $1.3 billion to keep it going by the year 2,000. The President feels this is a subsidy to beef and dairy producers that shouldn't be borne by the government. We see this a little bit differently—as a burden borne by the livestock industry, but if that view effects a change in appropriation, we're all for it.

One press release on the subject in its entirety states: "Pesky Disease Outlasts the Best USDA Could Throw. After 47 years and $2 billion the Agriculture Department is giving up its fight to eradicate brucellosis, a pesky disease that causes cattle to abort and that can infect humans. Budget officials said it doesn't make sense for USDA to spend $90 million a year to destroy a disease that caused $44 million in losses to the cattle industry last year.

"President Reagan has proposed slashing the $90 million budget by $31 million next year and lowering it to $6 million in 1985. Budget officials claim that cattlemen and state agencies will begin programs to make sure the disease is kept under control. Some USDA officials, however, fear brucellosis will become widespread again.

"Brucellosis is found in southern states, mostly Texas and Oklahoma.

126

Human beings who drink milk from contaminated cattle often develop high fevers. Fewer than 1,000 people suffered from brucellosis related fevers last year. USDA said about five herds out of every 1,000 contract the disease."

Our spies tell us that the reason this decision was reached is, in part, it has been found that brucellosis can never be eradicated because there is a vast reservoir of this disease in wild animals and, therefore, it is futile to continue with an eradication program that can never succeed. We find this conclusion a genuine revelation since this is the point we have been trying to drive home for years!

Control is the only real answer that has merit, and we have never been opposed to control. We firmly believe that, left to private enterprise, this disease will be controlled in short order, as it is to the benefit of the producer to do so. We also feel, that without government interference, control will be reached in a way that benefits the entire livestock industry. Now, it looks like we may have the opportunity to prove that theory.

BRUCELLOSIS CONCERN UNFOUNDED

Editor's Note: The following article covers the culmination of nearly a year of negotiation between the Federal Park Service (Dept. of Interior) and the state of Montana. The subject has been followed by *BUFFALO!* and reported in articles concerning the proposed brucellosis study on the transmission of the disease between wildlife (elk and buffalo) and domestic cattle in Montana. The author of this proposed study is a former federal veterinarian with the U.S. Park Service (Yellowstone) and is presently the Montana state vet.

Dr. Glosser has seen this problem from both sides of the fence, as a federal employee whose position was that the brucellosis infected elk and buffalo herds of Yellowstone posed no threat to nearby domestic cattle and now as a state vet who is pledged to protect the state's livestock from possible infection by these same infected herds. It is understandable that Dr. Glosser would propose a study to answer this question once and for all.

We find it very interesting that Wind Cave National Park research biologist Rich Klukas feels that this suggested threat is non-existent. We have been saying just that for years. We also find it interesting that he mentions there has never been a known contraction of this disease between elk and domestic livestock. This is also true between buffalo and domestic livestock as stated in Dr. Becton's presentation to the National Buffalo Association in Houston in 1979. We can't help but wonder, "why all the fuss, when it's never been shown there is a problem in the first place?"

Another interesting statement is Klukas' contention that stress (of the animal being tested) can alter the results of the test. This is a conclusion the NBA office came to quite some time ago and questioned Dr. Becton on. Dr. Becton admitted that there are many factors that can alter the results of the standard test and that stress very well might be one of them. He had no answers on how to deal with the variability of the results, however, except to keep them penned and keep testing. This, of course, increases the stress situation for a wild animal and further alters the results.

The most interesting fact we find in the article is that the National Park Service, with all of its budget and manpower resources, feels that a test and hold procedure for the elk would be impractical. Penning these animals for 30 to 90 days to determine that they are brucellosis free was too much to ask so they have decided to forget the whole thing and instead run the elk out of the Park and into the National Forest where hunting permits will be issued to reduce their numbers. Buffalo breeders have been up against this very thing for years and even though buffalo are wild animals and penning them for extended periods to test and retest is impractical, producers have been forced to do just that in order to stay in business.

We would like to stress one more time that the study proposed by Dr. Glosser and the state of Montana is a worthy project and needs our support. The only way this harrassment will ever come to an end is if a scientifically approved study determines there is no contraction of brucellosis between elk and buffalo and domestic livestock under normal conditions. In the meantime we will be labeled "guilty by circumstantial evidence." Here is the article:

Hot Springs—Excess elk problems are nothing new at Wind Cave National Park. Park Superintendent Les McClanahan says various attempts have been made over the years to control the elk population.

In the late 1950's into the 1960's, surplus elk were slaughtered and the meat sold in cooperation with Custer State Park. That effort ended after controversy erupted over Yellowstone National Park's "direct reduction" methods.

Then, with varying degrees of success, attempts were made to drive elk from Wind Cave into Custer State Park. McClanahan says some of the first helicopter drives were successful, but the elk got wise and refused to be driven. In the late 1970's, attempts to drive elk west onto national forest land for a state licensed hunt was "totally unsuccessful." The elk wouldn't be pushed through the heavy forest.

The Oglala Sioux meanwhile had asked to take elk for the Pine Ridge Reservation fee hunting program. Thirty elk were obtained in 1969 simply by closing the bison trap after the elk wandered in. A drive into the bison trap in 1970 produced 106 elk.

Jicarilla Apaches in New Mexico put up $10,000 to fix up the trap and a 1971 drive brought in 301 elk. In 1972, the Apaches took 200 elk, the Oglalas 153 and Custer State Park 126 in a cooperative roundup.

"We were pretty successful in getting the herd down to manageable size," says McClanahan. From a high of about 800 elk, the Wind Cave herd was trimmed to 400.

Since 1972, the Park Service has worked with several Indian tribes around the country to take elk. "The whole system was working well," McClanahan said. The tribes were paying the costs. The elk were being vaccinated and tested for brucellosis, with infected animals destroyed and only clean test elk shipped out. "There were no big problems," McClanahan says.

Last year another big roundup was set to take 250 elk from Wind Cave for tribes in Oklahoma and Montana. The helicopters were even on the scene ready to go.

But the Montana state veterinarian said no elk would be allowed into the state unless certified free of brucellosis. That would have required penning the elk 30 to 90 days and retesting until no infected animals were found. This was impractical and the roundup was canceled.

Park Research Biologist Rich Klukas suggests the concern over brucellosis at Wind Cave is unfounded. Of 614 elk tested since 1971, only 13 reacted for a "very low" 2.1 percent. In 1976, no elk reactors were found and in 1977 only one reactor turned up among 189 elk. Four reactors were found in 36 elk tested last year, but Klukas says stress can increase reaction to the brucellosis test and last year's sampling was done early, on a hot day.

Chief Ranger Dean Shilts says most brucellosis transmission problems are associated with crowded "feedlot" situations such as found on the Jackson Hole, Wyoming, elk wintering grounds.

McClanahan says it's questionable if brucellosis is transferred from elk to cattle. For that matter, there are many unanswered questions about brucellosis. Montana officials, after their ruling on admitting elk, are interested in doing research on the disease's transmission, he says.

VETERINARIAN URGES SHIFT IN BRUCELLOSIS EFFORTS

After 40 years of mixed results in brucellosis eradication efforts, it's time to shift the federal government's approach to the livestock disease problem, buffalo producers were told.

At the 15th annual fall meeting of the National Buffalo Association, Montana State Veterinarian, James Glosser, urged that the bulk of the federal brucellosis eradication fund, or $89 million this year, be focused on seriously infected Southern states, where much of the brucellosis invading relatively clean Northern states originates.

In the North, Glosser would leave it to state livestock sanitation agencies and economically motivated individual stockmen to control the disease.

Brucellosis causes cattle, buffalo, elk and certain other ungulate animals to abort and occurs in humans as undulant fever.

"Northern livestock producers, faced with loss of animals if brucellosis isn't controlled, will take it upon themselves to vaccinate against the disease," Glosser said. That's essentially what happened after Montana, several years ago, separated itself from the federal eradication program. (In South Dakota, new legislation not only makes the federal program automatically the state program but specifically provides that buffalo are to be treated as cattle.)

Glosser said, "Vaccination combined with strict refusal to import livestock from brucellosis infected areas has significantly reduced incidence of the disease in Montana."

"The federal program of the Animal & Plant Health Inspection Service, however, is based on blood testing animals and slaughtering those that carry brucellosis antibodies. This may help knock down high levels of infection," said Glosser, also a former federal veterinarian.

"Unfortunately," he said, "the test can also incorrectly indicate that vaccinated animals carry the disease. In states where the disease is virtually under control and preventive vaccination is widely used, this can result in excessive condemnation and slaughter of clean animals."

Glosser further said, "The federal program hasn't addressed problems of interstate shipment of livestock capable of transmitting brucellosis. And the national program does not meet the needs of the buffalo producer."

Buffalo raisers have been critical of federal brucellosis eradication rules on several counts. Especially they are concerned that the test and slaughter method can result in eradicating ranchers rather than the disease. They say they're also penalized by rules requiring successive blood testing because buffalo don't take kindly to corrals necessary to collect blood samples.

Glosser was one of several veterinarians, biologists, and buffalo producers who spent Friday morning gnawing over, and often disputing, the many-sided questions of brucellosis control with its special problems of buffalo herds.

COLUMNIST AGREES

We recently learned that we are not the only ones with serious reservations on the Federal Brucellosis Eradication Program. This is something we have long suspected, but couldn't document. Throughout the long history of this 'test and slaughter' program (approximately 40 years), there have been periods of rebellion and non-compliance from the cattle industry, but we have been led to believe by the federal veterinarians who have spoken to us at conventions the past few years that all is serene now, between the program and the industry. Not so.

Since we have gone public with our "battle," we have learned there are pockets of rebellion in many other areas. For the most part, these are unorganized movements in different areas of the country, but as groups of rebellion find each other, organizing begins and with organizing you develop strength. In recent months, we have made contact with several of these groups and expect more to surface as time goes by.

Be it buffalo or cattle, the one thing we all have in common is the belief that 'test and slaughter' will never eradicate brucellosis, the test is an unreliable rip-off and the program is designed not to eradicate brucellosis but to perpetuate the test and keep those federal vets gainfully employed.

The time is ripe, right now, for a major rebellion. The present program is cumbersome, costly, and ineffective. Livestock producers are getting tired of being pushed around and they are tired of taking it in the wallet when production costs are already out of sight. And the present climate is for austerity of budget on the federal and state levels. With all of these things in mind, we expect to see some changes in the near future.

We have tried to work through the proper channels (the bureaucratic officials of the program) to gain relief from the undue hardship this program places on the buffalo producer, but have met nothing but road blocks and double talk. Our recent meeting in Washington is a classic example of the run-around we have been getting.

That meeting did serve a purpose, however. It convinced us that we would get nowhere with gentlemanly tactics. The only route open for relief is political. Presently, there are several congressmen and senators from South Dakota and Iowa who are assisting us. But we do need political support and this is where you can help. We need you to contact your representatives and ask them to support our position on this "rip-off" program.

The following "On the Edge of Common Sense" columns by Baxter Black, DVM, state our position in a nutshell. Coincidently, we never visited with Dr. Black until after his first anti-brucellosis program column appeared, but he is saying just exactly what we have been saying for years. Through Dr. Black, we have made contacts with some of the leaders of the "Oklahoma Rebellion." Dr. Black tells us there are a lot more people out there with similar feelings.

Dr. Black is a syndicated writer whose columns appear in *The Record Stockman* (Denver, CO), *Livestock Weekly* (San Angelo, TX) and *Grass & Grain.*

ON THE EDGE OF COMMON SENSE

by Baxter Black, D.V.M.

COLUMN 1

The brucellosis eradication program has outlived its usefulness and should be quietly put in the burn barrel. I don't think brucellosis in the U.S. can ever be eradicated. The sooner the livestock industry admits this to ourselves the sooner we'll be able to get on with the business of controlling bangs like we control vibriosis, lepto, IBR, redwater, and a passel of other diseases that we control with proper vaccination.

Regular annual vaccination with a killed Brucella product (so there is no danger to the rancher) would certainly give as effective control as we have now. Under the present eradication program this is not allowed, mainly because it interferes with the "blood test." That is why the present Strain 19 live vaccine is not allowed in mature animals. Not because it wouldn't enhance protection against the disease, which it probably would, but because it interferes with the "blood test!" Seems like we've gotten the cart before the horse. It's like tryin' to get some cattle operation to fit the computer instead of makin' the computer fit the cattle. By the way, the killed Brucella vaccine is not available in the U.S.

This whole Bangs eradication program is a very debatable and lengthy problem which will surely be the subject of future columns. So now that you know my general opinion I'd like to comment on two areas:

1) The use of reduced dosage vaccine.

This is allowed in some states. Various state and federal agencies have decreed that qualified people may dilute the regular strength Bangs vaccine 100–200 times and use it. What bothers me is I've heard more than one regulatory veterinarian, while explaining why brucellosis is difficult to vaccinate for, say, "Even the Strain 19 vaccine is only 70 percent effective." Even without my diploma it doesn't make sense to me that reducing the dosage is going to give equal protection. Of course, if my main concern was the "blood test" then it does have a certain logic.

2) State laws allowing importation of Bangs vaccinated mature animals only.

This is popular in many legislatures right now. I have had occasion to testify against these proposed laws. They restrict trade, pure and simple. They do benefit breeders within a state since fewer cattle will be allowed to enter. But these laws play havoc with feedlot and cow-calf importing states. Think twice before you support these proposed laws and if your state passes them at least include a five or seven year end point on them.

Folks, don't think just because I'm against the eradication program I'd quit vaccinating heifer calves. Under the present laws you better vaccinate every potential replacement on the place!

Since I'm proposing eliminating the brucellosis eradication program which is written in stone in Washington, DC, I imagine some of you think I'm really standing on the edge of common sense! But if we're still trying to eradicate brucellosis in the year 2000 we'll know which edge.

COLUMN 2

Last month I presented my personal opinion that brucellosis was impossible to eradicate; that we ought to do away with the brucellosis eradication program and get on with the business of controlling the disease like we do vibrio and lepto.

There was a time in the '30's and '40's when Bangs was a serious abortion problem in beef and dairy operations. In 1934 the USDA instituted the eradication program. From '34 to '57 tremendous gains were made in controlling brucellosis. The percent of infected cattle tested in the U.S. went from greater than 10% to 1%. The federal program was largely responsible for this.

It was also a serious public health problem for people who worked around livestock and those who drank raw milk.

Now, you ask, how can I challenge such a "mother, home, and apple pie" program? First, eradication is a noble goal and served for years to reduce the incidence of this dread scourge on our cattle industry. The efforts were not in vain. But for the last 20 years we have spent countless tax dollars and put many a good man outta the cow business and not done much to reduce the incidence of brucellosis.

I think it is because it's impossible to eradicate and still have the cattle industry we have today! The simple fact that brucellosis has been found in and can infect swine, horses, dogs, sheep, goats, mink, buffalo, elk, man, rodents, and coyotes means that it is not just cattle that are carriers.

The crux of the matter is the difference between two words: ERADICATION and CONTROL.

Eradication is the goal of the federal program. It involves restricted cattle movement, restricted sale of breeding stock, slaughterhouse testing without the owners' permission, restricted vaccination, extensive blood testing, quarantines and slaughter of animals designated reactors. If you haven't paid much attention to the new set of brucellosis regulations that will soon go into effect, get ready 'cause you'll know it when they go in force!

Control, on the other hand, means vaccination for the prevention of disease. Once you eliminate the need to worry about the "blood titer," one of the killed Brucella vaccines can replace the Strain 19 vaccine. It can be given routinely to cow herds like we give vibrio or lepto. Two initial shots and annual boosters. The rancher can do it himself and quit worrying about Bangs.

Brucellosis is no longer a significant public health problem. Ranchers, veterinarians and packinghouse people run the greater risk.

We've been at this project for nearly 50 years. The last 20 years we're having trouble holdin' our ground. I'm just convinced we've done all we can under the present program and more government restrictions will wipe out a lot more cattlemen than brucellosis. I'll be 64 in the year 2009. That will the the 75th anniversary of the brucellosis eradication program. How many of you think it will be eradicated by then?

BRUCELLOSIS UPDATE

Final action was taken on the agricultural appropriations bill during the Dec., 1982 lame duck session. The measure was signed into law. As predicted, the conference between House and Senate committees "split the difference" on brucellosis program funding and agreed to a budget of $84,346,000 for fiscal year 1983. APHIS requested $92,000,000. USDA recommended $60,000,000. The "split difference" obviously was in favor of APHIS.

Members of the conference committee took great issue with USDA's recent change in the indemnity rate and included the following instruction with the appropriated funding:

"The conferees direct that the rate structure for brucellosis indemnity payments, which was in effect on Nov. 25, 1982 will continue. Further, the conferees direct the Animal and Plant Health Inspection Service to defer any revision to brucellosis indemnity payments until such proposals are submitted to the appropriate committees of Congress for their review in connection with the 1984 budget request.

"The conferees take note that both the House and Senate have strongly expressed their continuing support of the brucellosis eradication program. Consequently, the conferees intend that no reprogramming of brucellosis funds will be approved during fiscal year 1983, and that only a fair share of administrative absorption will be approved."

The Administration's plan was to phase the program out over a three-year period with $30,000,000 cuts annually to a projected skeleton surveillance budget of $7,000,000 in 1986. The conferees disagree with that philosophy as they have strongly expressed their "continuing support of the brucellosis eradication program."

There may be some hope on the horizon, however, as the funding bill contained $350,000 appropriated for synthetic vaccine research. And in March of 1983 APHIS was brought back in to testify before the House Agriculture Appropriations Sub-committee to justify the agency's handling of the appropriation. Committee members, though they re-appropriated the Brucellosis Eradication Program, are concerned with the direction the program has taken and the number of federal regulators on the payroll. They have determined that there is one regulator for every 100 infected animals in the United States. The House committee feels staff could be cut in half without jeopardizing the effectiveness of the program as a reward system to encourage compliance. The House committee is concerned that this program has become a punitive program with little incentive to comply.

Buffalo are now indemnified as a grade animal at $50 per head. Buffalo are also specifically mentioned throughout the rules, i.e., cattle and bison.

For years our basic points of contention have been that buffalo were subject to slaughter under the program but there were no provisions or considerations made for the unique handling and management problems that confront buffalo producers.

Now, it appears, buffalo are being dealt with even harsher than they have been in the past. Since buffalo are specifically mentioned in all the cattle provisions, there is even less flexibility under the program and harsher state regulations can be expected as a result. The major question we want to ask right now of APHIS is: "Is this program designed to protect cattle from potential losses from adjoining buffalo herds, or is it supposed to protect buffalo producers from the losses of brucellosis?" We need to know what they are trying to accomplish with their U. M. & R. which so frequently mention bison. If setting the buffalo industry up as the "boogy-man" threatening the cattle industry is their goal, then the present rules make sense. If the program is meant to protect our industry from economic loss and from transmission to other herds, then it is a blatant failure.

Buffalo cannot and will not be handled under cattle criteria. The economic loss factor as applied to cattle cannot be applied to buffalo either from the standpoint of herd management, longevity or economic loss because raising buffalo "is a whole different ball game than raising cattle."

If you want to know the rules you must live under in the future contact: USDA, APHIS, Federal Building, Room 848, Hyattsville, MD 20782 and ask for a copy of the Uniform Methods and Rules for Brucellosis Eradication. Be prepared for many pages of government "gobbledegook." The revisions take up about 30 pages alone. Read this carefully and then digest it as it applies to buffalo. That is the crucial part. If you are unhappy with what is being imposed upon you, contact your congressmen and senators. They ultimately have the final say on what comes out of USDA and goes back to APHIS. If nothing else, ask for independent consideration of buffalo under the program.

13
Parasite Control That Works!

A PROVEN PLAN
by Charlie Tucker

(Charles Tucker, 2-term NBA President, has been raising buffalo in the humid east, where land is high and space is tight, for a long time.)

The following applies to everyone who keeps buffalo on less than 10 acres per head—and to anyone else who sees any parasite symptoms.

I like the idea of buffalo spreading all over the continent, but those raisers with lush green pasture or two or three acres per head better plan on keeping up a very strict parasite control program or you are going to have problems. Not only that, but you could give buffalo a bad name. As is usually the case, people blame the animal when things go wrong, when it's probably their own fault.

When I started raising buffalo, there was no information on buffalo husbandry and my losses were staggering; they almost put me out of business. We've learned through hard experience and now have a healthy herd anyone could be proud of.

If you have only two buffalo, you can get away with putting Kibbles in their feed. Judging by the many telephone calls I get from NBA members, this is the system that gets most people into trouble. They tell me they've been worming their animals regularly and they just can't be down due to parasites. After a few questions, I find out they have been putting the wormer in the feed or water. Worming buffalo free-choice works great on the healthy ones, but it doesn't do a thing for the sick, timid ones. The only way to be sure they get what they need is to put it down their throats or inject them. Personally, I like oblets. Shoving it down the throat is a little harder than putting it in the water, but whoever saw anything good come easy?

You should give this forced treatment in the fall after the chance of parasite infestation is gone. If this is confusing, read up on the life cycle of a parasite; it may surprise you. It's far too long and complicated to explain here.

After their fall forced treatment, I follow up with phenothiazine in the salt in the spring. This is especially important in the spring if any of your animals look poor. Don't blame it on a "hard winter." You may have more of a problem than you imagine. If more than 1/20 of your herd is poorly, I highly recommend you worm them again before the grass greens. If nothing else, get to the sick ones.

Most important, when that grass starts growing, take away all salt blocks and minerals containing salt (and most do—read your labels). Replace them with a mixture of salt and phenothiazine (one part pheno to 12 parts salt). Mix well and keep this in front of them, free-choice, all summer until the parasites have gone dormant for the winter.

Phenothiazine is not going to take the place of tramisol oblets, it is merely a suppressant to keep the parasites in check until fall. I got into trouble because I thought if I wormed them with tramisol (lavamisole hydrochloride) in the spring, I was all set for the summer. I found out the hard way, this is not so in areas of moist grass. Those little buggers can build up enough in a buffalo between March and October to do him in!

Once the infestation period (March through October) is over, remove the pheno/salt mix and go back to your salt blocks or whatever else you use until the following spring when the cycle begins again. This switch helps to prevent a buildup of immunity to the phenothiazine.

DISEASES AND PARASITES

In the wild, buffaloes are hardy animals, comparatively free from diseases. But they are susceptible to a number of diseases that affect domestic cattle, and for that reason they should not be placed in pastures with domestic stock or otherwise be closely associated with them. Buffaloes are susceptible to hemorrhagic septicemia, blackleg, piroplamosis, anaplasmosis, brucellosis, necrotic stomatitis, and other infectious diseases of livestock. Immunity or increased resistance to many of these diseases may be afforded by vaccination with appropriate biological products. The services of a competent veterinarian should be enlisted for the prevention or treatment of disease.

Big game animals that are fed during the winter frequently become infected with necrotic stomatitis, commonly known as "sore mouth." The infection follows lacerations caused by coarse or harsh hay that permits entrance of the causative organisms. Young animals tend to become infected more frequently than adults. Prevention is easier than cure, and hay containing grasses having stiff, sharp awns, as foxtail and needlegrass, should be avoided.

Feeding repeatedly on the same ground allows unused feed, manure, and other waste to accumulate and furnish suitable breeding sites for the organism of necrotic stomatitis. The danger of this infection becoming established in herds under these conditions is greatly increased. A new place for feeding should be selected daily, and affected animals should be isolated if possible.

Many parasites that affect cattle also affect buffaloes. Ox warbles, the larvae of botflies of the family Oestridae, are frequently found in buffaloes, although heavy or troublesome infestations are not common. Buffaloes may become infested with ticks in localities in which these parasites occur to the extent that some of the animals may become affected with "tick paralysis."

Gnats and other flies cause considerable annoyance to buffaloes during the summer. Swarms of flies gather about the flanks of the animals and by their repeated biting, abraded areas may develop. Moist places where the animal may wallow and become coated with mud afford temporary relief from these pests. The screw worm fly creates another management problem. Any wound, even very minor ones, may become infected, and animals must be treated to prevent losses.

Internal parasites commonly found in buffaloes include lungworms (*Dictyocaulus hadweni* and *D. viviparus*), stomach worms (*Haemonchus contortus*), and intestinal worms (*Oesophagostomum radiatum*). Heavy infestations of lungworms may lead to a verminous form of pneumonia and cause death. Stomach and intestinal worms, when abundant, have a debilitating effect, causing emaciation and anemia. Young animals are most frequently affected and they seldom recover from severe attacks. Animals weakened by these parasites readily succumb to pneumonia or other infections.

Feed lots or similar areas in which the animals are concentrated for long periods furnish favorable conditions for the spread of internal parasites. Such areas should be eliminated from the range of the buffalo to reduce danger of infestation.

HERD IS NOW HEALTHY

"The Onondoga Nation's buffalo herd is now a very healthy herd due to direction given by many people such as Charlie Tucker & others.

"The buffalo were all badly parasitized. Ralston Purina's Cattle Ranger Wormer, a tetromisole, was used to worm the herd and it worked very well. It was fed to the herd at one 50 pound bag per 25,000 pounds of herd or approximately two pounds per 1,000 pounds of animal or two pounds per head. It was fed on the ground and readily eaten.

"The herd is now being wormed spring and late fall and is fed 16% grain from December 1 through April 15 along with a good timothy and alfalfa hay. 'WHAT A DIFFERENCE!' says Norman H. Goldstein, DVM, Manlius, NY, Veterinary Clinic."

E. L. Furnas of Terre Haute, IN, reports, "The morning the young heifer died, the herd came up the hill to where she was lying to visit her. I told my wife maybe they were going to get her up. She said, 'I hope so, but I think it's a wake.' She was right, because the calf died a few minutes later. Soon afterward, my herd bull, Old Ern, started losing weight and having diarrhea. By the time he had lost 200 pounds, I knew it was worms. We used two 10 pound bags of worm pellets, but I might as well have given him peanuts!

"About four weeks later, an old gentleman was telling me a story about how the Indians fed willow leaves for this kind of animal problem. Next day, I picked about 150 pounds of creek willow leaves. Old Ern ate them lying down. Before the day was over, he was standing and eating. Within a few hours the diarrhea stopped. In less than a week, he was grazing again. We keep willows convenient now, and no more problems."

PREVENTIVE ACTION

"We ration antibiotics," reports raiser Gary London. "Judging by what the rest of the lot is doing, based on the weather. We meet the stress of drastic weather change with preventive medication such as the mycins for about three days: AS 700 or aureomycin 338. Those are complete additives into the feeds. We use no implants."

London reports a trick he's learned on keeping flies down. "With chemical fly control, after three generations you've bred immunity into them. We kill flies with a wasp. The flies don't develop immunity to wasps.

"This is a flightless little wasp that looks like a flea. The wasps lay their eggs on the fly eggs and the wasp larvae eat the maggots. We get the wasp eggs from a Colorado insectory. We'd been spending $2,000 a year on chemical fly control. Biological control costs us only $1,000."

"Buffalo are exciting," sums up Gary. "They're a different type of animal. We want to get more experience. I've been 14 years in cattle with Alta Verde, but buffalo are unpredictable."

PARASITE CONTROL

The first step in getting your buffalo clean of worms is to have fecal tests run on fresh feces. Worms vary in every area.

Then sit down with your vet and discuss what product seems to work against this species of worms in your area. Different wormers are definitely more effective against different worms and the manufacturer does not draw this to your attention. The fact is some of the worst worms are hardly touched by some of the common hard sell wormers.

Three weeks after you worm you can run a fecal test and have them checked. Then you should be able to follow a program for sheep that is successful in your area, or you may want to continue with your veterinarian running fecals and prescribing a wormer for you once or twice a year. Twice is best if you have a low, wet pasture. –Kenneth Throlson, DVM

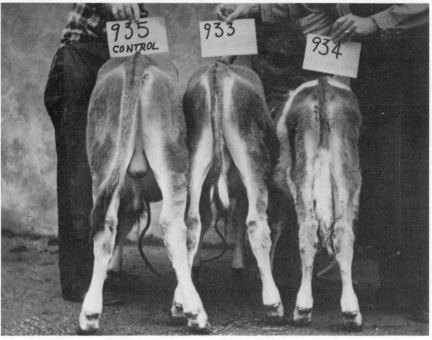

These calves are within four days of the same age. To show the effects of parasitism, 933 and 934 were inoculated with known numbers of worm larvae. Note the rough coat on 924 and the size differences compared with the worm-free control calf, 935.

141

The National Bison Range at Moise, MT, uses Spotton on all their buffalo. I have used Bo-Ana on a few of mine a number of years ago when confined in a small corral and it handled the problem. I used the cattle louse dosage which is half of the grub dosage.—Kenneth Throlson, DVM

DON'T BE BUFFALOED BY INTERNAL PARASITES

by Stanley E. Leland, Jr.

All of our wild and domesticated animals serve as hosts to a vast array of internal parasites. The "Monarch of the Plains" the American bison, or as we have dubbed him, the buffalo, is no exception. Our discussion here will address the internal parasites, and of these, only the worm parasites. Even within this limitation there are over 30 different kinds of worm parasites that are known to parasitize cattle and it is from the research in cattle that we are able to extrapolate to the situation in the bison. I will consider the parasites of the bison in three major groupings.

The Gastrointestinal Nematode Worm Parasites

There are over 25 different worm species in this group which are found as adults in the abomasum, small and large intestines. With sufficient numbers of worms present, clinical signs include severe diarrhea, emaciation, unthrifty coats, anemia, weakness and death.

Nearly all the worm species in this group are less than 1½ inches in length; some are barely visible with the unaided eye at necropsy. The adult worms are, therefore, not readily detected in the feces.

In order to first understand, and then to control these parasites, it will be necessary to have a working knowledge of their life cycle. With but few exceptions the species most commonly involved have similar life cycles. Five stages separated by four molts are involved. Two free-living stages occur outside the host on pasture or forage. The third stage (infective stage) is transitional in that it occurs on forage or in water and develops no further until ingested by the host. The total time required to reach the infective stage is approximately seven days under ideal conditions. The remaining two stages develop in the gastrointestinal tract of the host. The latter of these two stages consists of adult worms which, after mating, produce eggs. The eggs are passed out in the feces, hatch, and develop into the free-living stages, thus completing the cycle.

The type of environment determines in part the number of free-living stages that will develop and succeed in infecting the host. From the standpoint of necessary moisture, cover, temperature, etc., the irrigated pasture provides the best conditions for larval development, followed in order by regular pasture, drylot, and feedlot. Most bison have access to pasture and the infective larvae that may be present. The particular characteristics of the pasture also play a role in the transmission of infective larvae. Low partially shaded areas that retain soil moisture provide the ideal environment for survival of larvae. These same areas will likely also provide the most attractive forage and cause the bison to concentrate there. Thus, the bison pick up the infective larvae with their grazing and recontaminate the area with their egg-containing feces. This is nature's way of insuring that host and parasite get together.

High stocking rates will obviously lead to increased opportunity for infection and reinfection. In a case report from New York, clinical gastrointestinal parasitism including nine deaths occurred in a herd where the stocking rate was 30 bison/28.8 acres (0.96 acres/head).

Unthriftiness, weight loss, and death have been associated with cattle harboring large numbers of gastrointestinal nematode (round worm) parasites. On the other hand, cattle with small numbers of parasites may show no *readily* detectable ill effects (subclinical or inapparent parasitism). The delicate balance that exists between cattle and their worm parasites is incompletely understood but can easily be shifted in favor of either the host of the parasite, depending on therapeutic, nutritional, husbandry and hygienic practices used.

In obvious clinical parasitism, the course of action is usually clear but it is far more difficult to bring the clinical condition under control than it is to practice preventative management. These characteristics of parasitism are true for both cattle and bison. If anything, the accumulating information suggests bison are more susceptible to the effects of parasitism in general than cattle are.

LIFE CYCLE
GASTROINTESTINAL NEMATODES

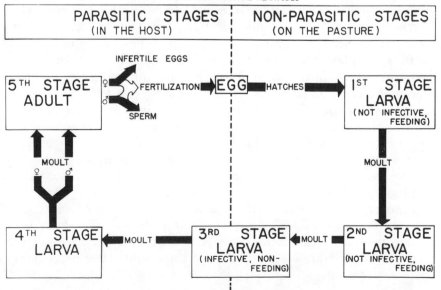

The Lung Worm

One or more species of lung worm are known to infect bison. One of these species is also found in cattle and deer. These are long white worms up to four inches in length. The adults are found in the trachea and bronchi of the lungs.

The lung worm life cycle, as in the case of the gastrointestinal worms, has five stages. There are some important differences, however. The eggs laid by the adult female lung worms are caught up, swallowed, and hatch in the intestinal tract of the host. Thus, larvae, not eggs, are passed out in the feces. Methods normally used to detect eggs in the feces for diagnostic purposes do not recover larvae and a special procedure is necessary where lung worm infection is suspected. The free-living lung worm stages are very susceptible to desiccation so high rainfall, humid areas of the country are where lung worms thrive. However, you should keep in mind that infected animals may be transported to other parts of the country.

The bison becomes infected when the larvae are ingested while foraging. The larvae penetrate the wall of the small intestine and migrate to the lung where they develop to adults, thus completing the life cycle.

Symptoms include those associated with pneumonia, increased respiratory rate, weakness. In severe or complicated infections, death may occur.

The Flukes

The Rumen Fluke and Liver Flukes can cause serious consequences in the American bison including marked weakness, intermittent and profuse diarrhea and death. These parasites have certain aspects of their life cycle in common which are important in management and control. The eggs that are passed out in the feces hatch into free-living forms. These forms swim in the surrounding water and must enter a species of water snail for further development to take place. When development in the snail is complete, the parasite leaves the snail and encysts on herbage or other objects in the water. Bison become infected by ingesting the encysted forms on herbage. In the bison, the parasite seeks out a particular site; the rumen flukes in the rumen, the liver flukes in the liver.

It is thus apparent that the cycling of fluke infection from animal to animal requires the introduction of the parasite, which is usually by way of infected sheep, cattle, or bison, a specific snail population, and pasture conditions suitable for snail survival and infection. The pasture conditions which influence the snail population in essence determines the geographical distribution or areas of prevalence for the parasite. Thus major areas of liver fluke exist in the Gulf states from Florida to Texas, and in the northwest states. However, small geographical areas exist in many states including Kansas. One must keep in mind that infected bison brought to any area can suffer the ill effects of parasitism even though transmission to other animals does not occur.

We have thus considered three major groupings of internal worm parasites, the gastrointestinal nematodes, the lung worms, and the flukes. We now know how the bison becomes infected and what signs or symptoms suggest parasitic disease. We need now to consider what possible management practices or options are available to counter parasite buildup both within the bison and in his environment.

In bison showing clinical symptoms of parasitic disease and with diagnosis confirmed where possible, by appropriate laboratory tests, treatment with an anthelmintic drug will be necessary. Your practicing veterinarian is your best source of assistance in establishing the diagnosis and selecting the appropriate medication. A number of compounds are available that are suitable for bison. Some of these medications can be placed in the feed. Others are injected or introduced orally.

As we have seen, all these major parasites spend part of their life cycle outside the bison either free-living on the pasture or in the snail intermediate host. It is at this point in the parasite's life cycle that you, as a producer, can exert an important influence in breaking this cycle.

Reflect with me for a moment over your own particular management practices and pasture or animal holding facilities:

- Are you aiding transmission by overstocking?
- Is pasture rotation possible within your management plan?
- Are young more susceptible animals in contact with older, more resistant but infected animals?
- Do you know what the level of parasitism is in your herd? Have you had fecal examinations?
- Do you have a treatment program for the cows calculated to minimize exposure to the susceptible calves?
- Do you isolate newly acquired stock to determine their parasite load and need for treatment before allowing them to contaminate your pastures?
- Do you have low lying pasture areas that are poorly drained and have standing water?
- Do you know the source of new stock with regard to geographical origin and what parasites they might have and whether or not they have been treated?
- Within practical limits do you manage to separate feed from accumulated feces?

Finally, my discussion is not intended to unduly alarm or frighten you. As man has proceeded into the process of animal production, he has, as is the case with the bison, taken an animal with comparably unlimited confinement and both concentrated and restricted it. This action, which is necessitated by the demands of society and economics, favors disease transmission and particularly the parasitic diseases. There are, however, many examples of success with parasite control in the pasture confinement process and this should occur with bison if you don't let yourself be buffaloed by the worm parasites.

KENTUCKIAN HAS GOOD RESULTS WITH PARASITE, FLY CONTROL

by Bill Austin
Kentucky Buffalo Park
Mammoth Onyx Cave
Horse Cave, KY 42749

We have gotten good control over internal and external parasites. We started them on MoorMan's IGR Minerals prior to the fly season. This contains 0.02% Methoprene (Isoprophy) (E.E) -methozy 3, 7, 11, trimethyl -2, 4-dodecadienoate tradenamed ALTOSID. This chemical acts as a birth control pill for the horn flies and prevents their breeding in the chips.

We also hung up a Cow Life-Cattle Rub between one of their favorite utility poles with a guy wire and treated it with four gallons of diesel laced with Shell's CIOVAP for face fly control. Contrary to our expectations, they took to it and have really enjoyed using it. Amazingly, the device has survived their affections to date. We have seen very few flies on or about the animals.

After running them last December (for brucellosis-free certification), we wintered the survivors on rough hay supplemented by alfalfa and 50 pounds of MoorMan's Mintrate Cattle Cubes every second day for the 19 head. We renovated 25 acres of bison show paddock in the winter and offered them 50 pounds of MoorMan's Cattle Cube type wormer before moving them on the new ground.

One two-year-old bull that had gotten through the chute in December before we caught him looked a bit off condition still, so we loaded him up with an injectible wormer via Red Palmer's wonderful device. He now looks as good as the rest. We've gotten good service from the Cap-Chur system from the first attempt and believe it has saved a few animals for us. I think the most important use of these sorts of devices is to medicate the animals without having to capture them.

A good example of that is our 1982 calf crop arrived in force just nine to ten months after we worked the herd bull over with the dart rifle and 15cc syringe-darts loaded with injectible wormer.

14
Tranquilizers and Vet Procedures

CAPTURE AND RESTRAINT OF BUFFALO
by Nelwyn Stone, BS, MS, DVM
and Terry Rettig, BS, MAT, DVM

This is meant to give the layman a base working knowledge of buffalo capture and restraint. In addition, the technical data on equipment, drugs and dosages are generally unavailable to most veterinarians when the information is needed. Technical data are included so that the buffalo owner can either use them himself or have them available when others come to work with his animals.

The most commonly used capture equipment in the U.S. today is the Cap-Chur brand manufactured by the Palmer Chemical and Equipment Company of Douglasville, Georgia. Another company is Pax-Arms of New Zealand.

Since the Cap-Chur equipment is the most widely used, it will serve as our example. There are three types of projectors which can fire a syringe-dart into a buffalo. The short range "pistol" propels a syringe using compressed carbon dioxide. Its range is 13.7 meters, or 40 feet. It can be used on buffalo at close range, especially when they are closely confined. The strength of the CO_2 charge varies widely with temperature, and may be quite weak in cold weather.

The long range "rifle" is also powered by compressed CO_2 and has a range of 32 meters, or 100 feet. Again the propellant charge may be very weak in cold weather. The extra-long range "rifle" is powered by a .22 blank charge and has a range of up to 80 meters, or 250 feet. The power of the charge does not vary with temperature, making this projector more suitable for cold climates.

Cap-Chur syringes vary from one to 15 milliliters with needles from 1 to 5 mm long. For buffalo we recommend the NCL needle with a 3.75-cm length and 3 mm outside diameter. A barb located about half-

147

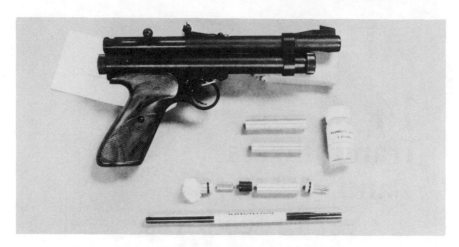

Short range gas powered projector and darts.

Long range gas powered projector and darts.

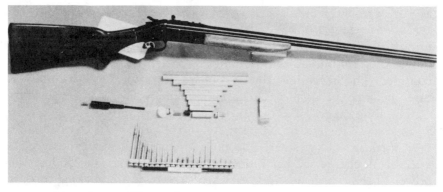

Extra long range powder powered projector and dart syringes.

Palmer Co. photos

way down the shaft of the needle grips the skin long enough to allow the drug to be injected by the powder charge contained in the syringe. Colored tailpieces help locate the syringes and help identify syringe contents. Further information may be obtained from the "Catalog of and Instructions for Use of Cap-Chur Equipment" printed by Palmer Chemical and Equipment Co.

A buffalo's neck, shoulder and hindquarters are well fleshed and make good targets for drug injection. The brisket may seem a good target, but a buffalo tends to lower its head when the projector is fired and a syringe may hit the animal in the head. If possible, never tranquilize a buffalo near a body of water. Drugs and excitement may cause a buffalo to overheat. He will seek out water and may drown or become hopelessly mired. It is best to confine the animal to a small enclosure, preferably padded. However, if you could do that you probably wouldn't need to tranquilize him.

Buffalo are ruminants. Vomiting (regurgitation) occurs easily and a drugged buffalo may choke. If possible, it's a good idea to fast buffalo at least 24 hours before immobilization. Keep the animal on his brisket with head elevated. A buffalo lying on its side cannot burp and may bloat rapidly. Keep the animal cool, using shade or water sprays. A drugged buffalo's eyes will dilate and direct exposure to the sun can cause blindness. Cover the eyes to prevent damage.

DRUGS

Atropine Sulfate. Dose—up to 0.04 mg per kg body weight. Duration—less than one hour.

Atropine decreases salivation, stomach secretions, sweating, and intestinal motion. Atropine increases the heart rate and causes the pupils to widen. Atropine helps reduce secretions caused by Ketamine or pheneyclidine. When using atropine, protect the animal from heat and his eyes from light.

Acepromazine Maleate. Dose—up to 0.1 mg/kg body weight. Duration—up to one hour.

Acepromazine is a potent tranquilizer which depresses the nervous system and reduces the dosage of anesthetic needed. "Ace" can be used with etorphine, tylazine, ketamine, or phencyclidine. It prevents vomiting, reduces blood pressure, relaxes muscles, and calms aggressive animals. Because it can slow the heart rate, it should be used with atropine. It can be given IV (in the vein), IM (in the meat) or orally (by mouth). The effects of oral administration can be unpredictable, however.

Xylazine (Rompun). Dose—sedation, 0.1–0.3 mg/kg; anesthesia, 0.6–2.0 mg/kg. Duration—two to four hours.

Xylazine can be used with local anesthetics such as lidocaine or with general anesthetics such as chloral hydrate, barbiturates, or halothane (chloral hydrate and barbiturates can only be given IV and halothane is a gas, so therefore will not be discussed as capture and restraint drugs). Xylazine can cause a decrease in heart rate, respiratory rate, and blood pressure. It should be used with atropine. The animal should not be disturbed until full effect has occurred as excitement may prevent the drug taking effect. Animals can still kick while under Rompun 0.2-1.25 mg/kg and Etorphine, total dose of 3-6 mg.

Fentanyl/Droperidol Citrate (Innovar-Vet). Dosage—up to 0.1 mg/kg of Fentanyl. Duration—less than one hour.

Droperidol makes the animal indifferent, and reduces movement. Fentanyl is a potent short-acting pain killer. Innovar-Vet can be used for short minor procedures.

Etorphine (M99, Etorphine). Dosage—0.004-0.002 mg/kg. Duration—less than one hour.

Etorphine is a highly potent anesthetic. It is very dangerous to humans as a small amount absorbed through the skin or mucous membranes can be fatal. Etorphine increases blood pressure and heart rate and decreases respiration, intestine motion, and body temperature.

Diprenorphine (M50-50). Dosage—about twice the given dose of Etorphine. Duration—less than one hour.

The effects of etorphine can be reversed with diprenorphine.

Nalorphine. Dosage—0.25 mg/kg.

Naloxone. Dosage—0.006 mg/kg.

Nalorphine can also be used to reverse the effect of diprenorphine. In humans, diprenorphine, nalorphine, or naloxone can be used intravenously in case of accidental exposure to etorphine.

Etorphine can be used in combination with Xylazine or Acepromazine.

Tiletamine Hydrochloride/Zolazepam Hydrochlorise (Tilazol) together are known as CI-744, which may be marketed soon. CI-744 can be used as a sedative or a general anesthetic. The eyelids may remain open and many of the reflexes remain functional. Excessive salivation can occur but can be controlled with atropine.

Few of the chemical restraint drugs listed are approved by the Food and Drug Administration for use in other than domestic animals. Most are potent and dangerous drugs which cannot be obtained without a prescription from a licensed veterinarian. Some are classified as controlled substances by the U.S. Drug Enforcement Administration. Some are easily absorbed through the skin, giving the unpleasant side effect of death. Accidental injections may also give this undesirable side effect. For current regulations concerning the use of immobilizing agents, contact your regional office of the Drug Enforcement Administration.

References:

Fowler, M. E. 1978. *Restraint and Handling of Wild and Domestic*

Animals. Iowa State University Press, Ames.

Fowler, M. E., ed. 1978. *Zoo and Wild Animal Medicine.* W. B. Saunders Company, Philadelphia.

Harthhorn, A. M. 1976. *The Chemical Capture of Animals.* Bailliere Tindall, London.

Klas, H. G. and E. M. Lang. 1982. *Handbook of Zoo Medicine.* Van Nostrand Reinhold Company, New York.

Young, E., ed. 1975. *The Capture and Care of Wild Animals.* Ralph Curtis Books, Hollywood, Florida.

BUFFALO: CAN THEY BE TRANQUILIZED?

by Paul Seabold
Seabold's Animal Farm
Valley Road, R. R. #1
Keokuk, IA 52632

Editor's Note: The following is Mr. Seabold's presentation on tranquilizing buffalo, which was presented at the NBA's fall meeting in Moberly, Missouri, on September 12, 1980. The presentation and ensuing question and answer period have been edited for clarity, and are based on his first-hand experiences, not hard and fast rules.

Seabold: "Tranquilizing guns are not guns. They are dart projectors. There are two types, short range projectors and long range projectors. The short range projector can be used even in the pen, up to three feet away with no injury to the animal. At this close range, the animal can be medicated with a drop-off needle. This is an easy way to medicate without harrassing or getting in close confinement with the buffalo. The long range projector has a range of up to 300 feet. If you can't get any closer than that, you're not going to treat the animal anyway.

"I use a tranquilizer in the feed the night before I want to handle the animal. (PROMAZINE GRANULES) There is no danger if they get too much so I give four times the dose on the label. I use ROMPUN and ACE PROMAZINE in the dart. The animal is tolerant of both drugs. If he gets too much he'll just sleep a little longer. If the animal goes down and sleeps, be sure to lay him on his left side with his head hanging down to eliminate the animal strangling on his own fluid. But, basically, its a very tolerant drug. I use DOPRAM-V to level out their breathing (if they go down). ATROPINE SULPHATE is used to stop slobbering (salivating). The two are good to have on hand in the event a tranquilized animal gets an overdose. But the tranquilizers I use give good results with no bad reaction. ROMPUN is the best you can get ahold of in the opinion of many vets."

Q: Is there an antidote?

Seabold: The antidotes are DOPRAM-V and ATROPINE SULPHATE. Dosage, depending on the weight of the animal, approximately

1cc ROMPUN mixed with 2cc ACEPROMAZINE (used in the dart).

Q: If an animal died as a result of these drugs or improper handling, would the meat be edible?

Seabold: No it wouldn't. This is a bad thing about it, but the drug will not kill them. He'd have to run into the fence and break his neck or something, so it's not likely you'll have that problem. You're not going to tranquilize him like they show on "Wild Kingdom," where he's all laid out. All you're going to do is take the edge off him so he's not going to crash around and break a horn in the corrals or injure himself.

Q: If you used the drugs, ROMPUN and ACEPROMAZINE to settle the animal for hauling to the packing house, when would the meat be edible (safe for slaughter)?

Seabold: It'll wear off in 12 to 24 hours. So if you drug him at night, load him up and then haul him to the packing house the next morning and they kill him at noon, it'll probably be alright.

Q: Why is it that a lot of times, when a vet tranquilizes a buffalo, it dies?

Seabold: That's because in most cases the animal is excited, keyed up, hot. They're using it as a last resort. If an animal in our area gets out, we'll tranquilize him first. We won't go chasing him with pickups, snowmobiles, and motorcycles. We'll get him calmed down and then handle him. That's why a lot of vets have adverse results. They do it backwards. Too many people use tranquilizing as a last resort. It should be the first thing you turn to.

Q: What's the dosage on the drug used after they're tranquilized?

Seabold: The one to level out the breathing is DOPRAM-V. It comes in a 20cc bottle and I give a flat 1cc in any sized animal. The other one (to stop the salivation) is ATROPINE SULPHATE. I give a flat 1½cc to anything. Calf, bull, or cow. It dries them up.

Q: Do you use a shallow injection?

Seabold: I shoot that in the muscle.

"Another reason I use the short needle, which goes against what a lot of vets will say (where the needle just goes under the skin), is because the reaction is more predictable. If you're trying to shoot in a muscle and you happen to get into an artery or vein and it gets into the blood stream quickly, that's where they drop faster. This way (short needle) is more predictable to me."

Q: DOPRAM-V and ATROPINE SULFATE (the antidotes), will that be a factor in the meat if the animal dies?

Seabold: No, it would not be. You could use the meat if there wasn't enough tranquilization in them otherwise. But if the animal goes clear down, always lay it with its head down and preferably on its left side.

Q: Do you use ACEPROMAZINE and ROMPUN together?

Seabold: We use PROMAZINE GRANULES in the feed to take the edge off; then later if they need to be settled down some more, shoot them with ROMPUN. The dry feed won't put them out near as much as when you dart them.

Q: What are the chances of bloat?

Seabold: None whatsoever as far as any tests have shown so far.

Q: Some tranquilizers cause bloat, don't they?

Seabold: It's not the tranquilizer that causes bloat. It's the buildup of fluid and the difficulty with breathing and the gasping for air that causes the problems.

Q: Have you used and can you use all of them together? PROMAZINE in the feed, ACEPROMAZINE and ROMPUN in the dart later?

Seabold: Yes, if I've left the feed for them and then the next day don't think they've gotten enough, I've used the dart on them with the other drugs with no adverse effects.

Q: How much would you use?

Seabold: Same dosage. The dosage I give with the liquid. We've worked it out very well. It makes the animal drunk, he hangs his head and he's not going to charge you and you can maneuver him on his own power. We try to keep them where they can maneuver on their own power. Dosage depends on the weight of the animal. Approximately 1cc ROMPUN mixed with 2cc ACEPROMAZINE (used in the dart).

Q: After you've tranquilized them to move them and they're down on their left side with their head down, where do you go from there?

Seabold: If it happens to go clear down on me, if it's a sick animal or something, we load them up with an unloader or we get slings under them with enough man power to just slide him into the trailer. My trailer is just nine inches above the ground. After we get them in the trailer, we try to lay them on their left side. Then I get in my truck and drive like hell to get where I'm going.

Q: Do you sling them under the body?

Seabold: No. When I say sling, I mean like a stretcher. We lay them on it to get them on the trailer. We just slide them up.

Q: Our vet lays them (cattle) on their left side, but keeps their head up. If it goes down, he tries to get it back up. He says he does this to avoid bloat. What you're saying is just the opposite.

Seabold: I wouldn't let him do that if he was working for me. I wouldn't let him handle anything, even a mouse that way. A vet will tell you ROMPUN is made for horses, but if you write the manufacturer of the drug they'll send you dosages for all kinds of animals and it's working well for my buffalo. But they're (the manufacturer) still working on their recommended dosage for buffalo. They don't want to say you can use their product in conjunction with another manufacturer's product (as I do). That's a bug-a-boo with them. So they haven't got around to establishing a dosage for buffalo yet.

153

Q: I understand that buffalo are, perhaps, the most difficult animal to tranquilize. Is this true?

Seabold: I haven't found them that difficult. In the variety of animals that we handle, I haven't found any more adverse results from handling buffalo than anything else.

Q: What are the problems with buffalo, that you feel would require tranquilization?

Seabold: Well, one thing, they chase me . . . My prime use for it is where I acquire a lot of animals from zoos or individuals. They're raised as a pet, but for one reason or another, the people have to get rid of them. They're not going to let anyone come in and shoot it for meat. That's where I come in.

Q: Why would you want or need to tranquilize?

Seabold: For moving them, to load them, to get one down and check his feet, for any number of reasons that you had to get in close contact with him. When he's tranquilized he's safer and easier to handle.

Jerrell Shepherd: "From a breeder's viewpoint, let me answer that. When a buffalo gets sick, normally he either gets well or he dies. There's not much you can do about it. That's one of the problems with raising buffalo. The larger, western type ranches probably never know when they have a sick animal as they are out on the open range. They don't know till they find a dead one. The ones that recover on their own, they never know about.

"When you're raising these animals under eastern conditions, they're a little more likely to get sick and less likely to recover unassisted. If you're not careful, you're likely to have worm problems, you're going to have real problems out there. And if a buffalo gets injured and you need to treat him and you don't have corral facilities to handle him, you have more problems. Even with facilities, oftentimes your sick buffalo is not near your facilities. One nice thing with tranquilization is that to some degree you can handle these animals and treat them.

"Most small operators begin their operation without enough fencing. They think they can contain them (buffalo) with normal fencing and they can for awhile . . . till they decide to roam. Then they'll jump the fence or go through it. They don't have any corral facilities at all, they don't have a squeeze chute, they don't have any way at all to treat an animal. By the proper use of tranquilization, as Paul has been discussing, you can get an animal to where you can treat him.

"If you just call a normal vet out and ask him to tranquilize an animal, I'd say nine times out of ten you're going to lose that animal. I think that tranquilization has a great part with the small buffalo breeder that has not yet built the proper handling facilities. That's my point of view as a breeder. Now, Paul is an exotic animal dealer and he has to go out and load these animals under the most adverse circumstances and he tranquilizes more animals in a month than most of us will in a lifetime and he's had so much experience at it that I felt he was the one who could answer our questions."

Q: On field loading, say, you've fed the animal PROMAZINE GRANULES the night before and you're now going into a reasonably small enclosure or pasture, do you use those little portable metal corral sections?

Seabold: Yes, six foot gate, six feet high, and I try to get the animal, if he's real tranquil, into a far corner, then ease the trailer up as close to him as possible. Don't get the animal excited. Just sort of squeeze up. Another thing that helps facilitate loading, I've found, is if you can load in the evening. If you put a light in the front of the trailer, they'll run to the front of the trailer.

EXPERIENCE WITH CHEMICAL TRANQUILIZERS

H. C. "Red" Palmer probably has more experience than anybody handling buffalo with dart administered tranquilizers. He is president of Palmer Chemical & Equipment Company in Douglasville, GA. Palmer manufactures gas and powder powered projectors. The big ones look like shotguns, the little one like a pistol. Compressed gas or powder charge launches a dart syringe that injects, upon contact, 1 to 15cc of liquid into the animal.

He reports that, after several drugs had admirably proved themselves over 22 years, the FDA banned them. As of this writing, only ROMPUN may be used.

Palmer advises, "ROMPUN can be a great help in handling your animals. I recommend that you experiment with your equipment and ROMPUN, starting with a low dosage and shooting your animal from time to time, gradually increasing the dosage to the point where you can treat the animal without danger of losing it.

"The dose of ROMPUN depends on WEIGHT, and your animal's PERSONALITY—how wild it is. Naturally, the wilder, the more drug it will require, since it is a sedative plus analgesic. With little effort, the animal can be stimulated into action when you approach or start to handle it. Remember, always be careful for your own health's sake."

This is important enough to bear repetition. A certain dosage may induce the animal to lie down and be quiet—yet leave him capable of sticking horns into you when you come within range. Red points out that you do not always have to knock the animal out. You may be able simply to calm it to the point that it can be roped, trussed, and treated without undue exertion.

When working a bunch of buffalo, one technique is to shoot them

You don't always have to knock them out. Sometimes it is sufficient to quiet an animal so it can be roped and stretched without a township rodeo.

all with enough sedative to quiet them so you can cut one out or come in and treat him without the others bothering you. Be prudent when tranquilizing your animals. Red relates that one young bull was accidentally injected twice and went down, although the rest were only quieted. The owners insisted to tailing him up and getting him onto the truck with the others. But the unfortunate animal went down in transit. His friends trampled him to death. These sorts of things must be kept in the back of your headbone when using chemical control.

Red says, "Any drugs that will immobilize a buffalo cause increased body temperature. If you are going to immobilize a buffalo in summer, do it in the cool of the morning or late afternoon. Watch that temperature! If he goes down and stays on the ground for a time, he will get extremely hot and can die of heat stroke. If he remains down for a time, roll him across the brisket from side to side frequently to dispel bloat and to prevent leg paralysis.

"Use care with ROMPUN," cautions Red. "Even if the animal is not completely immobilized. When he is left alone, he will lie down and go to sleep. If you don't make him get up often, he may die of bloat though he never was completely immobilized."

ELECTRONIC TRANQUILIZING

Tranquilizing buffalo with drugs has advantages over the lasso—sometimes. But tranquilizing buffalo is such a sometime thing, any improvement over drugs would be welcome.

Now from Australia comes an electronic gadget, the Stockstil, that Bill Moore, Bar X Ranch, up north of Cody, WY, likes real well. About the size and shape of the six-volt lantern, it's powered by the same lantern battery and so is totally portable. It has two needle electrodes.

Once you have an animal in the chute, you insert one electrode under the tail and the other into the cheek and turn it on. The animal stiffens and stays motionless and silent while being branded, castrated, dehorned,

With the aid of the electronic tranquilizer, Bill milks a buffalo cow.

or whatever. The animal returns to normal within seconds after you turn the switch off.

The manufacturer says the device "in no way electrocutes the animal. The current is only 1/15 that needed to light a small penlight." You can hold the electrodes in your hands and not feel a thing. It passes 1-millisecond pulses of 240 MA current through the animal, not enough to hurt, but enough to contract the skeletal muscles, thus immobilizing the critter. Muscle contracture "produces a pain-blocking effect which reduces, if not completely eliminates, any pain," he claims. He said Australian stress

Bill Moore with the Stockstil.

One electrode is inserted under the tail, the other in the cheek.

level tests showed stress was "several times less than that experienced by cattle that went through painful procedures such as dehorning without being immobilized.

Bill Moore even milked a buffalo cow who was plugged into the device! He thinks the procedure is not uncomfortable for the animal, because he'd been giving this same buffalo cow daily treatments with it in the chute and, "She didn't become chute shy, as she would have if she was hurt," he observed.

An article in a farm magazine quoted an Iowa vet who uses it for dehorning cattle, castrating calves, and performing Caesareans on gilts. He felt it took a while to learn how to use it best.

The veterinarian thinks the contracted muscles also clamp off blood vessels and reduce bleeding. He reports the device reduces stress not only in the cattle—but in him and his assistant, saving them many kicks and scrapes. "It's faster than tranquilizing an animal and waiting for it to take effect."

Kenneth Thorlson, DVM, and NBA buffalo rancher of New Rockford, ND, observed Bill Moore's demonstration. He takes violent exception to electronic tranquilizing, which further illustrates the fluid state of the buffalo art: We don't know much for sure about buffalo—there's still lots of room for contrasting opinions and new knowledge.

States Dr. Throlson, "I flat out don't like it. I think they feel it but are powerless to move or bawl. If you put the front lead on the forelimb or behind it, the animal will bawl and bawl. All ruminant animals are much more sensitive to electricity than man." He can cite many professional observations supporting this view.

15
One Vet's View: Buffalo and Me

by Kenneth Throlson, DVM
New Rockford, ND

I have been intrigued with animals all my life and especially with buffalo. It seemed the more you left them alone, the better they fared. I purchased some land to build a veterinary clinic on and with this there was 25 acres of hayland, so I started planning a fence and looking for buffalo. In November, 1978, I purchased my first buffalo and have raised them since. Any statements I make in this article are strictly my opinion and in no way would I consider myself an absolute authority on buffalo.

In 1981, I sold my practice and set out to follow a dream of buffalo ranching. I bought 320 acres, plan to run 80 buffalo cows on that when I'm through developing it; 268 acres were cropland. I worked about 200 acres of it and seeded it with a plane to 50% alfalfa, 25% bromus and 25% intermediate wheatgrass. This is mostly Type II soil; rainfall averages 18 inches. I've built up to 76 head (about 45 cows) and would like to get to 80 cows on this half section. I take a lot of my winter hay from highway ditches and sloughs.

I work part time for a beef feedlot in Carrinton, ND, setting up the health programs and seeing that they are carried out. My wife and I have six children, so we need the cash flow of this job while developing the ranch. I enjoy this very much and it helps me keep up with newer things in my profession. I also sell some hay each year and do some custom baling.

For the first six years I sold all my calves. Then last year, having acquired the added acreage, I started saving some of them.

In '82 I started marketing some meat. I butchered six and am building my meat market as I go. Response has been very good.

I have raised buffalo for nine years now and have observed them for hours and hours (being late for meals and events). The animal almost has me under a spell. After a successful roundup I said to a friend, "If we both raise buffalo until we are 100, we may just begin to know something about the nature of the beast."

When clients asked me what I would like to do if I quit practice, I answered, "Raise buffalo, since you don't have to walk around with a syringe giving antibiotics to them all the time."

IMMUNITIES

Which brings up immunities. If buffalo are allowed to run over a large area, I believe they have a native instinct to protect themselves from diseases, parasites, malnutrition, etc. However, when we confine them, we impose upon them a limited range for protecting themselves. Also, as some writers say, we may in some way break that intrinsic factor of their spirit—we have all seen men so crushed by bankruptcy, divorce or a death that they soon succumbed to some disease. Most, if not all, the susceptibilities of buffalo have been given to them by man. We in science often feel that we have made tremendous strides in bettering the world, but the buffalo has come all the way essentially on his own. I, for one, feel that we must be very careful in domesticating one of the last wild animals of real value to man.

We must be careful not to destroy him or make him completely dependent upon us, as we have made the cow. There is not another animal in the U.S. who has the capacity to turn wasteland into edible protein with no help from man. Being a veterinarian, I just cannot see myself enjoying a soybean or seaweed steak because of protein shortages.

Buffalo can be kept healthy if you do not confine them in too small an area, and give them a large choice of nutrition with minerals and salt available. I can run at least three buffalo where I would run two cows, but then I may need to start managing them as cattle. If I run the same number of buffalo as I would cattle, I don't feel I need to do much managing, other than providing trace minerals and salt which they cannot find in a limited area. Also, if you are in an area with specific soil deficiencies (i.e., the Great Lakes region soil is deficient in iodine) or toxicities (i.e., some arid Western soils are high in selenium) you may need to take care of this as you would for cattle or sheep.

BUFFALO ARE A LOT LIKE SHEEP

Buffalo are very much like cattle in the fact that they are ruminants, but some of their behavior and weaknesses in captivity may be more like those of sheep. I recommend you get your buffalo clean of worms, then follow a program similar to a successful sheep rancher in your area since each area differs in internal parasites. Check with your friendly corner veterinarian.

Buffalo do not require shedding as cattle do. In fact, if you build them a nice shed or barn, they won't go in. They should not be confined to an enclosed area. Next time you butcher a buffalo, check the size of the windpipe and compare that with the biggest beef you can find. If the

162

size of the windpipe has any bearing, they need lots and lots of fresh air. This may also account for part of their running ability. Also, the hump with the leverage of its vertebral spines, like a row of ginpoles, helps with the running and the great strength in the head and neck.

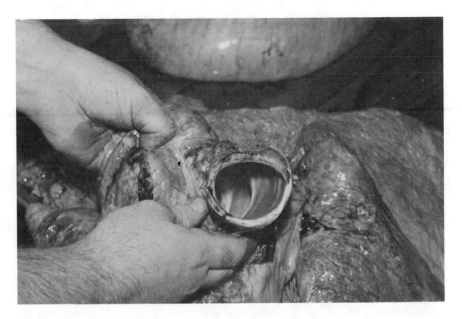

Huge trachea admits two to three times as much air as a cow's windpipe.

FUN & GAMES

Buffalo remain playful to a much greater degree than cattle. Cattle mature out in a year or two but 20-year-old buffalo are almost as apt to play as buffalo calves. There are days when the whole herd plays in the morning or toward evening.

The games my buffalo play vary. When it's cold, they play a lot of tag or follow the leader. There are about six acres of abandoned gravel pit in my pasture; they will all chase up and down and around those dunes for half an hour or so. Other times they appear to play chicken— they charge at each other at an angle and just miss; then turn around and charge again. When it's warmer, they like to butt heads. I have seen 1,100 pound cows fake a fall when calves that are so young they're still red, butt heads with them.

Buffalo roll, and I am sure most of us would think that cattle were crazy if they acted that way. It should be easier for cattle to roll, since they do not have the hump, but rolling is pretty well limited to buffalo and horses.

As a species, they are more of a social animal than even cattle. Cattle often scatter when chased or frightened, but buffalo clump—one reason the white hide hunters could wipe out a whole herd in a single stand. I have seen buffalo pushed through a fence and turn right around and fight to get back through the fence to be rounded up with the herd. This is probably why crossfences separating herds to not work well.

PROTECTIVE FEEDING INSTINCTS

They do not overeat as cattle do when they have access to something they like a lot. However, a few will die from enterotoxemia if fed too much high energy feed too fast. But I have never seen any buffalo bloat as cattle do from alfalfa or grain.

They definitely prefer coarser and finer grasses than do cattle. Whereas cattle prefer alfalfa and other legumes, my buffalo eat these only after they have eaten the slough grass, reed canary grass, June grass, and brome grass out of that pasture. Yet in the winter, when I feed alfalfa hay, they will eat it fine for a few days and then go and eat every spear of a slough bale.

One dry year, I wintered my few head on basically cattail bales. Last year, when my cattle didn't like what was left in the hayracks, I pulled the leavings into the buffalo pasture and they cleaned them right out—leaving the good bales alone. I feel we could fatten both beef and buffalo on the same pasture with a rotation system and not decrease the number of units of either by more than, say 25%, giving a larger return on the same acreage. My pasture is all domestic grass. The cattle would eat the alfalfa, and the buffalo the other grasses. But then we would need more management in the health care of the buffalo and defeat one of the reasons some of us like to raise buffalo.

THEY TAKE CARE OF THEIR CALVES

Buffalo are more likely to fight to protect their calves than cattle are. Also, buffalo calves nurse much more frequently than beef calves and they start to eat forage on Day One, which would be unusual for beef calves.

A buffalo will root through snow to grass and will paw through thicker ice than a horse will to get water, whereas most cattle give up and wait for you to give them water or feed.

BUYING TIPS

When buying buffalo at a public auction, be very selective. Know where they are from. Get a history of their health and management. Select individuals with nice hair coats, bright eyes, and alert behavior. Those with dirty, uneven and miscolored haircoats are often sick, will become sick or have nutritional problems showing up. Even cattle hair often bleaches out and becomes uneven with heavy parasite loads and nutritional deficiencies.

Hooves and horns can also tell a little history. Heavy rings on either show periods of prosperity and poverty. Broken and crumpled horns may show they have been through the chute often.

When buying at the ranch, ask to see the whole herd, not just those offered. Look at the general condition of every animal in the herd and make mental note of the calf crop, yearlings, etc. Ask about vaccination history and other procedures such as worming, any herd health problems, etc. Then pick the nicest haired, brightest eyed individuals along the type of conformation you want. I prefer the deepest colored, longest, largest animals; someone else may like the lighter colored, shorter, more humped buffalo. Remember, in breeding stock it's hard to breed champions starting with the bottom end.

STATE HEALTH OFFICES

STATE	Telephone Number	STATE	Telephone Number
Alabama	205-832-3760	Nebraska	402-471-2351
Alaska	907-745-3236	Nevada	702-784-6401
Arizona	602-255-4293	New Hampshire	603-271-2404
Arkansas	501-371-1311	New Jersey	609-292-3965
California	916-445-4191	New Mexico	505-247-2254
Colorado	303-861-1823	New York	518-457-3502
Connecticut	203-566-4616	No. Carolina	919-733-7601
Delaware	302-736-4821	No. Dakota	701-224-2655
Florida	904-488-7747	Ohio	614-866-6361
Georgia	404-656-3667	Oklahoma	405-521-3891
Hawaii	808-487-5765	Oregon	503-378-4710
Idaho	208-334-3256	Pennsylvania	717-783-5301
Illinois	217-782-4944	Puerto Rico	809-722-1159
Indiana	317-232-1330	Rhode Island	401-277-3047
Iowa	515-281-5547	So. Carolina	803-788-2260
Kansas	913-296-2326	So. Dakota	605-773-3321
Kentucky	502-564-3956	Tennessee	615-741-1441
Louisiana	504-342-4984	Texas	512-475-6488
Maine	207-289-3701	Utah	801-533-5421
Maryland	301-454-3831	Vermont	802-828-2421
Massachusetts	617-727-3015	Virginia	804-786-2481
Michigan	517-373-1077	Washington	206-753-5040
Minnesota	612-296-2967	West Virginia	304-348-2214
Mississippi	601-354-6089	Wisconsin	608-266-3481
Missouri	314-751-3377	Wyoming	307-777-7517
Montana	406-449-2976		

STATE LAWS

Each state has a state veterinarian who can give you the requirements for getting your buffalo home. (See accompanying list of telephone numbers where the desired information is available from a state health officer.) Please follow the law because not only are you protecting other buffalo and cattle raisers, but you may be protecting your own herd and finances. Illegal movement may result in a quarantine, fine, and even compulsory slaughter.

QUARANTINE ALL NEW STOCK

Now that I have corrals, when I bring new stock home, the first thing I do is put them through the chute and identify them (tag and try to age by mouth and horn), worm them, and vaccinate them. I use 7-Way clostridium, and brucellosis on heifer calves. These new animals I keep apart from the herd for three weeks and observe them daily to see how they are adjusting to their new feed, environment, and climate.

The first two weeks, I keep the newcomers separated from the herd by a 100-foot corral. The last week, I move them into the corral with just a single corral fence separating them. The first two weeks give me a good idea on the health of the new buffalo and prove if they are going to break out with any disease due to shipping. The last week gets them acquainted with the others and, even though separated by a sturdy fence, they do establish their pecking order to some degree. This lets me put new animals into the herd with no chasing and very little fighting. Even large bulls have established the peck order without fighting.

While confining these new animals, have the feces checked for internal parasites. Confinement gives you enough time to have blood tests run and get the results back.

LEAVE 'EM ALONE!

Generally speaking, the more you leave buffalo alone, the better they will do if they have adequate browse, salt, minerals, and water.

NUTRITION

They are foragers. They will pick out and eat what a beef cow will walk right by. They will go and eat where it is too much trouble for a beef cow to go. They also seem to do better as to haircoat, etc., when they are grazing as opposed to being fed. They shed their winter robe much earlier when allowed to graze. Most of us have seen zoo animals that look hideous even though they may be on a better worming and health program than we have for our animals. The reason is they are not allowed to graze.

In winter, mine receive a mixture of slough, brome, wild, and alfalfa hay. It seems the more variety you give them, the better they balance

166

it. However, I always sell my best hay and feed my buffalo what around here we call horse hay.

A buffalo and a horse will both live without water if there is snow, but they definitely do better if provided fresh water. They will open water for themselves through the ice if there are banks so they can get down to it.

HABITAT

They face into the wind and they need a lot of fresh, dry air. A few trees or breaks in the terrain is all they need to winter in North Dakota. They need enough space so the bottom ones in the peck order can eat and still stay clear of the others. Their area should be large enough that it does not get muddy. They actually get along better in rough terrain where it is difficult for us to walk than they will in mud.

Feeding should be done in such manner that feces and urine do not contaminate feed or water. Remember, buffalo eat in an established order, meaning calves are often eating what is dropped or pulled out of the rack or bale by an adult.

VACCINATION

I vaccinate my buffalo only for the 7 colostridiums which includes the overeating or enterotoxemia clostridiums.

The females I also vaccinate for brucellosis. Now that they are using the new diluted vaccines, we do not seem to have to fear vaccination titers that give false reactors or suspects such as occurred under the old system. I feel that, as an industry, we must all get together and use the tools available, even if they are not 100%, to protect ourselves and other buffalo ranchers and stockmen in the area. However, I feel the Federal government must also clean up its herds if it expects us private herd owners to be obligated to do so.

Some areas need to vaccinate for Lepto. This your local veterinarian can recommend. Or there may be still another disease in your area that he can help you prevent through vaccination.

SPECIFIC CONDITIONS

Cuts and Wounds

Most cuts seem to heal on buffalo as well by themselves as they do with suturing. If the wound is clean and the animal is on clean grass and flies are not a problem, it should heal nicely. When there is infection, clean it, give long acting penicillin, protect it from flies, and turn the animal out.

Pinkeye

If not treated, you will get some white (blind) eyes. Give 1cc of zaiumycin injected under the upper lid and spray with a topical antibiotic,

167

"For treating pinkeye, I prefer the patches that let the animal see out the bottom half."

and glue on an eye patch. After glue is dry, release the animal and forget about him. I prefer patches that let the animal see out the bottom half. The patch will fall off in two or three weeks—sooner in brush.

Often buffalo injure their eyes from rolling in sand, or they may get stickers in their eyes from weeds, etc. Treat these the same as pinkeye.

The most trouble-free method to prevent pinkeye is to reduce flies. Some use insecticide ear tags. Others feed rabon in the salt during the fly season. This really helps if all the livestock within, say, a mile are handled in the same manner.

Hoofrot

You can use penicillin and spanbolets to treat hoofrot. (Although buffalo seem to have resistance to hoofrot.)

If you are having a problem with this, your local veterinarian can provide you with iodine to mix in the salt (the level will vary with the area). This will greatly reduce the incidence of it.

Worms

I like to use an injectible wormer in my buffalo—it does a good job on intestinal and lung worms. However, it will not touch tapes or flukes. If you have flukes, you will have to work with your state vet to get permission to use an experimental drug. If you have tapes, you may want to use telmin drench. It not only takes out the tapes but it also does a nice job on the other intestinal and stomach worms.

Your vet can best advise you as to what wormer to use after doing a fecal exam for you. You can worm buffalo with boluses, drench, injectable, or drugs in drinking water or feed.

"I went against tradition and left lots of daylight through all my pens and runways."

FACILITIES

Early in your ranching you need to decide if you are going to have just a few pets or if you are going after the breeding stock as well as the meat market. If you plan to have any numbers at all (say 10) and want to sell live animals, you will have to build some strong corrals and chutes. You need them not only to handle the animals, but to protect yourself as well. I have seen corrals made from all different kinds of materials and most have worked to a degree.

In building my corrals, I went against tradition in that I left lots of daylight through all my pens and runways. My chute is in the center and I can see every animal in my pen from any location in the corral. But this type has to be stronger because buffalo will hit it harder if they can see through it.

I used 7 1¼-inch tube steel welded onto 2⅞-inch oil pipe stem posts 68 inches high on 8-foot centers. So far this has worked well. My reason for building this way was twofold:

1) I can see every animal.

2) It does not hold snow, and in this part of the country we handle the buffalo only in winter.

I use a plain cattle chute to hold the animals. For catching them, I made a gate that swings across the runway 18 inches in front of the chute, giving me plenty of time to close the headcatch on them.

You may build your corral out of whatever materials you can get with the least effort and expense—used pipe, used lumber, used railroad ties, all work fine. The more daylight you give the buffalo, the stronger you will have to build because they'll hit it harder. The more you close out light, generally, the less strength you need. But whatever you use, you will need very strong stuff, well-anchored where you put your chute.

Also with buffalo, the larger the groups you are handling, the stronger you need to build, since often the front ones can't stop because the ones behind are pushing them.

Covering their eyes usually stops them cold.

TECHNIQUES

Jacket Blinder

I use a cattle chute to brand buffalo. We throw a jacket over their heads as a blinder, take one wrap on each horn with the sleeves and lay the iron on them. Covering their eyes usually stops them cold. Usually

170

they do not move even when being branded. The jacket can be used for other procedures, too.

Dehorning

We use a guillotine dehorner and follow this with a hot iron to sear and stop the bleeding.

Castrating

From all the information I have, there is no merit in castrating. Even

Buffalo are easier to pill than cattle.

taste test panels could not detect which meat was which. Probably the only merit would be in having an animal that you felt safer with. If I were to castrate, I would use a knife and emasculator.

Hoof Trimming

Buffalo allowed any area to roam in seldom need a hoof trim. I have hung up one foot in the chute and trimmed it when there was an injury— again covering the eyes helps.

Balling

We use a nose lead and the jacket trick in giving boluses to buffalo. They're easier to pill than cattle.

Shots

Same technique as cattle.

YOUR VETERINARIAN AND YOU

How you use your vet depends on what kind of herd you are trying to maintain, what kind of relationship you have with him, and if he is one of us crazy vets who like buffalo. Call your vet when you first see an animal is sick or going off condition. Often chronics of various disease appear similar. "A dying animal is a dying animal." A few weeks or even months before that, he was a diseased animal. You may want to use your vet for an autopsy in the case of sudden death.

When you call your vet, have the animal up and sorted in a pen. Do not expect your vet to show more nerve with the animal than you, the owner, do. He may not have had experience with buffalo and all men are afraid of what they know nothing about. Also make available to him any information on the condition your animal is showing. Even if he has no first hand knowledge of buffalo, he is most likely better informed in disease processes and medication than the rancher.

FEEDING

In drylot feeding buffalo, be sure they never run out of water. A buffalo can be put on almost any kind of feed (he will do better if it is a balanced ruminant ration) but he does need a constant supply of water. If you are going to push them on a high-energy ration, vaccinate for enterotoxemia (clostridium).

Also allow plenty of room in the lot and with twice the bunk space per head that you'd allow for cattle. They do not eat shoulder to shoulder as cattle do. Some will not eat for fear of being hooked.

DO'S

DO enjoy your buffalo. Take time to study them and see how they differ from other animals.

DO provide for their nutritional needs, no matter where you raise buffalo. (Pretty much the same as cattle, only less.)

DO respect them. They have the quickness and strength to kill you without meaning you any harm.

DO pick the best animals when buying stock.

DO butcher the less desirable individuals.

DO exhibit and promote them.

DO get help when having a problem. A vet can suggest which medicine may be most effective for your problem. Remember, buffalo can be medicated in the feed, water, or injected. Often the feed/water method does the job without the stress and danger of handling.

DO eat buffalo.

DON'T

DON'T put them into a small pen and then not provide feed, minerals, salt, and water.

DON'T ever trust them—they can kill you.

DON'T think you can buy cheapies and turn them into winners.

DON'T think you can raise and market them without facilities that will hold them.

DON'T think you will get rich overnight raising buffalo. You need to be a good producer to make money with them, and it takes time.

DON'T keep poor stock around—cripples, chronics, etc., do the industry and you no good; they give yourself and other buffalo raisers a bad name.

DON'T overgraze or overcrowd them.

DON'T raise one buffalo alone. A buffalo and a horse are social animals. To do really well, there must be at least two. In some cases, a horse may substitute a person, especially a child, for another horse, but I have never seen one buffalo do well alone.

Marketing and Promotional Strategies

IV
IV
IV
IV
IV
IV
IV
IV
IV
IV
IV
IV
IV

16
Sell
on the Hoof
or on the Hook?

If you're into buffalo for a living, that ol' Bottom Line tells the quality of your life. If you're doing it for fun, that's the line that tells you if you can have more fun.

Selling buffalo is a mite different from selling cattle. There's no price established every day in some far off market that you have no control over. At this point, there is no central marketplace. The NBA is working on solving that quandary, however. You don't haul your animals to the buyer and just take any little thing he wants to give you. The buyer comes to you, hat in hand, and says in respectful tones, "How much do you want for your buffalo, please—Sir?" And HE pays the freight!

Buffalo, alone of all agricultural products, let you set your own price! This isn't to say that you can get a million bucks a pound, your price has to be within reason—but the reason is there's more demand for buffalo than supply. This gives you, as they say on Wall Street, "leverage"! It also means you may have to find, or develop, your own market.

ON THE HOOF

Let's say you have breeding stock, or live slaughter animals, about ready for sale. The most obvious way to drum up trade, outside of selling to your friends and relations, is to advertise. Classified ads are the economical and safe way to go. The most famous ad of all time was a half inch HELP WANTED ad in the London *Times* back about 1902. It promised low pay, long hours, danger and discomfort. It also drew hundreds of applicants—for an expedition to the South Pole!

The basic principle of advertising is constancy: Don't blow your roll on one big ad. Decide how much to invest in advertising, then run smaller ads regularly over a period of time. Obvious advertising media are *BUFFALO!* Magazine, your local paper, your state farm paper, and perhaps regional and national farm magazines if your budget allows.

Then there's the metropolitan press. A lot of gentleman farmers have businesses in the city—and a farm within a few hours' drive. They read the local metro daily. Also consider business magazines.

If you're big enough, you can have your own auction. Pray for good weather. If there are several buffalo raisers nearby, pool your sale stock for a bigger auction, bigger ads, a bigger crowd, and a bigger net. Rent the local sale barn.

On pricing, basically, ask what the market will bear. If you can't get more for your buffalo than for cattle, you might as well quit. Breeding stock should bring at least as much as good purebred beef. Slaughter animals should fetch 30 to 50% more than that day's market on beeves of similar age and finish.

CANADIAN IMPORT REQUIREMENTS

To export buffalo into Canada, you must follow the guidelines for beef. *Required tests:*
1) TB
2) Brucellosis
3) Anaplasmosis
4) Blue Tongue

You must have a clean test on every animal in the shipment lot on all four tests. They must be retested 30 days later and still test clean. If all are clean on all eight tests, you can move them within 30 days of the second test. If even one animal reacts on any of the required tests, none may enter the country.

You must then start over with the four tests 30 days apart. This sometimes goes on for months as one animal will react on one test and be clean on subsequent tests—only to have another animal reacting on the next tests, etc., etc., ad nauseam. It helps to have a good relationship with the Canadian government. It also helps to develop friendships with border crossing officials.

As soon as you determine what exactly they want, then you can have the forms filled out to their requirements. If you really want to ship to Canada, and are willing to suffer some problems on the initial shipment as you develop some simpatico, you will find it a lucrative market. There is now a Canadian Buffalo Association and NBA works closely with CBA. NBA can put you in touch with the CBA and can also provide a list of Canadian producers.

ON THE HOOK

Late fall and winter is usually the best time to butcher. The pelage is prime then, and fetches the best price. As was pointed out at the Cody meeting in 1982, buffalo yield the highest quality eatin' meat at age 3½, and they make the fastest, cheapest gains at that age. They keep on growing to age seven or eight, but gains are neither as fast nor as cheap and the meat begins to lose its peak of perfection.

BUFFALO INSPECTION: State laws vary on the time, place, and manner of inspection and depend on whether you intend to consume all the meat at your own table, sell it within your state, or go interstate. Some insist that the live animal be inspected. The following information should be helpful.

MEAT INSPECTION GUIDELINES

The Federal Meat Inspection Act identifies cattle, swine, goats and equine as species which must be mandatorily inspected by USDA before the meat of these animals is permitted to enter into commerce. Appropriated funds are made available by Congress each year to cover the cost of the mandatory inspection program.

The meat from other animals such as buffalo or reindeer is subject to the jurisdiction of state of local authorities; provided, however, that if the meat enters interstate commerce, it becomes subject to the jurisdiction of the Federal Food and Drug Administration (FDA). The FDA and USDA do not conduct a mandatory inspection program for buffalo meat.

At the request of the industry, USDA does provide a voluntary inspection program for animals not subject to the federal meat inspection act. The voluntary inspection programs are authorized under the Agricultural Act of 1946. They are performed on a fee basis with the recipient paying the costs at rates specified by the secretary of agriculture. The 1982 rate is set at $13.38 per hour per inspector.

At the time the regulations and procedures for voluntary inspection of buffalo were established the regulations for domestic species were applied, as buffalo were a large bovine animal similar in many respects to cattle. This means that the full range of ante-mortem and post-mortem inspection procedures would be conducted within the structure of the approved slaughtering facility. Ante-mortem inspection must be conducted at the point of slaughter. The inspected buffalo meat may be sold as such, or used with other inspected ingredients to make sausages or other processed meat food products. This meat, or further processed product, is eligible to bear the USDA inspection mark and comes under the jurisdiction of USDA.

In most instances the same procedures apply for state inspection of buffalo meat in states which conduct a state inspection program and have inspection laws governing buffalo meat intended for commerce.

For purposes of interstate commerce, inspection is not required by USDA. State inspected product and uninspected product is free to move interstate, provided the state of destination will accept it (currently about half will, a quarter won't and the rest aren't sure), the state of origin will

179

allow it and the ultimate consumer doesn't require it.

At the present time, there are 27 points in the United States which are approved to provide voluntary buffalo inspection services and apply the USDA stamp. Twenty-five of these plants conduct slaughtering, or slaughtering and processing operations, and the remaining two conduct only further processing operations. Buffalo meat, or meat food products containing buffalo meat, which emanate from these plants will carry the USDA inspection legend with establishment numbers between 5,000 and 5,049. These plants are also approved to conduct the slaughter of domesticated species, or further processing of meat therefrom, and thus are also subject to the mandatory program. During the 1981 calendar year, 2,471 buffalo were slaughtered under the USDA voluntary inspection program.

We estimate that between 8,000 and 10,000 buffalo went to slaughter in 1981. This leads us to the obvious conclusion that the majority of buffalo slaughtered in the United States is slaughtered either with no inspection or under state inspection programs.

The USDA approved plants as of 1982 are:
- Hynes Packing Co. (Slaughter), Paramount, California
- Kamery's Wholesale Meats (Slaughter & Processing), Olean, New York
- Karl Ehmer, Inc., Farm (Slaughter & Processing), Lagrangevill, New York
- Eastern Oregon Meat Co. (Slaughter), Baker, Oregon
- Beehive Machinery, Inc. (Processing), Sandy, Utah
- Hamilton Packing Co. (Slaughter & Processing), Hamilton, Montana
- Black Hills Packing Co. (Slaughter), Rapid City, South Dakota
- Northwest Data Services, Inc. (Slaughter), Kalispell, Montana
- Carson Valley Meat Co. (Slaughter), Minden, Nevada
- Stillwater Packing Co. (Slaughter), Columbus, Montana
- E. A. Miller & Sons (Slaughter), Hyrum, Utah
- John R. Daily, Inc. (Slaughter), Missoula, Montana
- Stanko Packing Co. (Slaughter), Gering, Nebraska
- Quality Packing Co. (Slaughter), Sterling, Colorado
- Big Country Meat (Slaughter), Craig, Colorado
- Steamboat Packing Co., Steamboat Springs, Colorado
- Peterson's Lockers, Inc. (Slaughter & Processing), Gothenburg, Nebraska
- The Hillsdale Packing Co., Inc. (Slaughter), Hillsdale, New York
- White's Wholesale Meats (Slaughter & Processing), Ronan, Montana
- Rise Meat Packing Co. (Slaughter), Oakdale, California
- Link Bros, Inc. (Slaughter), Minong, Wisconsin
- Vernon Meatland, Inc. (Slaughter), Los Angeles, California

- Rasmussen Meats (Slaughter & Processing), Missoula, Montana
- J. M. Bostwick & Son, Inc. (Slaughter), Caledonia, New York
- Tolman Meat Processing (Slaughter & Processing), Hamilton, Montana
- East Fishkill Provision, Inc. (Processing), Hopewell Junction, New York
- Durham Meat Co., Inc. (Processing), San Jose, California

The addition of beef fat to buffalo burger is not a USDA requirement, but is done at the option of the processor. A label "With Beef Fat Added" contiguous to the product name is required whenever the level of added beef is less than 20%. When the level of added beef fat exceeds 20%, the term beef fat must be part of the product name. In all circumstances, USDA does not permit the total fat content of a product labeled as Buffalo Burger to exceed 30%. This is the identical requirement which is in place for hamburger or ground beef.

Since buffalo go not as lambs to the slaughter, field slaughter is usually the method of choice to minimize damage to truck, buffalo and you. This, however, is not legal in all states and does not qualify for the USDA stamp of approval. In 1983, NBA formally requested that USDA review and revise the regulations under which buffalo qualify for the USDA stamp. NBA requested that buffalo slaughtered in the field or somewhere besides the approved facility, in certain instances qualify for the federal legend. NBA also requested that state inspected facilities be allowed to apply for and use the special USDA buffalo stamp. The buffalo meat inspection policy revision request was still being studied at the time this book went to print. NBA can update you on this matter.

For now the method of choice by most small producers is field slaughter. It makes your job easier if you can keep the animal off feed overnight, since an empty gut is easier to handle than a full one. The meat is better eating if you avoid frightening or exciting the animal. Ideally, you mosey up to him and put a 30-30 or .357 magnum into an ear. Try to pick soft ground, since meat can be bruised and damaged where he falls on himself.

It's important to cut the main artery in the throat fast, so that the heart will pump out all possible blood. Then get him up by the heels— fast. A front loader is the quickest way to hoist a ton or more. It takes a high reach. A big bull may stretch 14 feet when hung by the heels.

Gut him there. Then haul the field dressed carcass to farmyard or locker plant or wherever you're going to skin him and cut him up.

A PACKER'S ADVICE TO BUFFALO MEAT PRODUCERS

by Jon R. Howard, General Manager
Black Hills Packing Company
Rapid City, South Dakota

First, establish in your own mind the market you are processing for. The full service processing plant can give you exactly the right product for your market: Carcass, boned, boxed, vacuum packed animal cuts or subprimal cuts, fresh or frozen.

They can ship for you to any place in the country. You save freight by shipping full truckloads. Once the live weight is established, your packer can estimate the yield. On the hook, approximately 70 carcasses load a truck; boxed it takes 90 to 100.

If at all possible, have your buffalo accustomed to being penned together before loading so they won't be so abusive to one another. Don't overload them. Use straight trucks if possible instead of pots. Load and deliver to the slaughter plant no earlier than the previous night. It's better to load and deliver the morning of slaughter. We have adequate pens for them. However, leaving buffalo close penned for an extended time is not a good idea, they're very hard on each other.

In fact, it is best that all buffalo destined for slaughter be dehorned as calves. Dehorning is much kinder to the buffalo and much more profitable to the raiser, because all the time our people spend trimming away bruises is costly to him. Not only that, but the meat trimmed away comes right off the top of his profit. Besides, bruises lower carcass quality and hide quality and therefore price, all a dead loss to the producer.

I've seen buffalo with 30 or 40 puncture wounds—sometimes even punctured lungs. The hides sometimes look like pincushions from goring. Dehorn your calves—unless you want to market heads or skulls—it makes the carcasses and hides more salable because of less horn damage.

BHPC is a federally inspected plant with a capacity of 500 head per day. We have five or six major customers for whom we custom process buffalo—2,000 or 3,000 head a year, out of 6,000 to 8,000 that go to slaughter nationally per annum. I presume this makes BHPC the biggest buffalo processor in the world.

The captive bolt used to kill cattle won't work on buffalos' thick skulls, so we shoot them with a rifle just behind the horn into the brain. This does no external damage to the head.

We can run about 50 cattle an hour with one man killing. We need four men on the killing floor with buffalo, and the number we can process is considerably less, because buffalo are so much wilder, stronger, tougher, and harder to handle than cattle. Killing is charged by the head. Processing is charged by the pound. The minimum order for custom killing is usually 10 head.

We return green heads to you if you have a market for heads or skulls; otherwise they go with the rest of the inedibles into inedible rendering. The hide is ready for you the next day, salt cured and drained by raceway cure. This gives the hide a long shelf life. You thus are not pressed to get it further processed immediately.

Butchering buffalo is heavy work, harder than cattle. The heavy hair is tough to work through. But our people have done it for years. They are professional butchers and professionals do a better job. They attend to details, like leaving the switch on the tail; that's important in your hide marketing.

I advise buffalo meat raisers to choose a federally inspected commercial slaughterer and processor who is experienced with buffalo. This is

BHPC cattle handling facilities are designed for buffalo. Circular pens with center pivot gates make it easy and safe to shunt animals from pen to pen to chute.

The breaking line. Primal cuts are trimmed, packaged and frozen for delivery to customers.

183

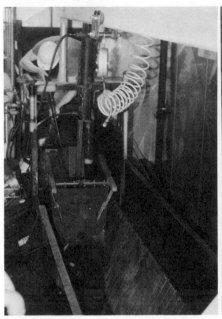

Pneumatic holder secures buffalo's head while rifleman centers a .22 magnum behind horns.

Eviscerating.

Splitting with pneumatic saw.

Organs slowly ride the conveyor belt past the inspector.

184

far better than field dressing. We recommend federal inspection even though it is not required on buffalo in that most of your customers are more comfortable selling or serving federally inspected products.

INNOVATIVE MERCHANDISING

Most operators will fill orders from the local locker plant. Some are big enough to have a walk-in freezer right on the home place. In either case, you have to let people know you've got good prime buffalo meat for sale.

ADVERTISING: One method is to notify possible markets within your area that you shortly will have prime buffalo available and, on advance orders, you'll cut and package to specifications. Prime prospects are supermarkets, hotels, restaurants, clubs, gourmet food shops, natural food stores. (Some will say "nay" because they don't want to stock a commodity unless they can depend on a steady supply. Others will capitalize on a one time buffalo feature for the publicity. Some clubs like to put on a big annual bash.) Whether you write letters, print circulars, make phone calls or go see the people personally depends on many things: How big a supply you have; how many potential outlets in how big an area do you want to cover, and your personal style.

BE YOUR OWN OUTLET: NBA former president, Lloyd Wonderlich, and his family set up a tent each year, hire a professional barbecue chef, and sell buffalo sandwiches at the Old Threshers Reunion in nearby Mt. Pleasant, IA. This celebration gets a quarter million visitors over the Labor Day weekend. At last count the Wonderlichs sold 7,000 ¼-pound chopped BBQ buffalo sandwiches at $1.50 apiece! They make a lot of friends for buffalo meat that way and they sell a good many frozen roasts and steaks and patty packages each year. And there's lots of room for applied psychology in merchandising. Lloyd discovered that "chopped buffalo" sells better than buffalo burgers.

Then there's the old hot dog stand. A little imagination can conjure up lots of ways to sell this most salable of all foods. NBA's 1982 President, Charlie Tucker, is friends with Ben and Marie Kassirer in his hometown of Stormville, NY. As a retirement activity, the K's opened a hot dog stand on Route 52 in Stormville. Charlie suggested adding buffalo burgers to their menu. Business perked right up. "Construction workers looking for a quick, nutritious lunch were the first to bite the bison," Mr. K reports.

Tucker says, "The K's are filling a void. Theirs is the first place you can buy a buffalo burger this side of Wall Drug Store in South Dakota."

Charlie ran out of burger a couple of months before the season's end. This taught the K's a lesson. They promptly ordered a year's supply for the next year because, with the passing of their buffalo feature, hot dog sales fell off dramatically. Charlie makes five patties to the pound. He gets $2.15 a pound and the K's sell them for $1.50 each.

The K's got a bonus out of the deal. Not only did they gross 300% on the buffalo, but the delicious and novel menu item drew extra customers from all over. The only promoting they did was to put a few signs along the road and they had about all the business they wanted until the buffalo burgers ran out, Charlie reported.

BE CREATIVE: Ways to profit from buffalo are limited only by your imagination. Why not raffle one off? Specialty food products like jerky, braunschweiger, sausage, etc., bring fancy prices. One fella gets $20 a pound for buffalo jerky flavored with fruit juices! Scraps of buffalo leather bring $4 a square foot (see the adjacent chapter on selling hunting privileges).

AIR EXPRESS: If you're near enough a jetport that you can offer to air express frozen meat, you might consider a small ad in magazines that appeal to the jet set, such as in-flight magazines, gourmet, and high style publications. For ad costs and sample copies write to the "Advertising Manager" in care of the magazine. Remember to make the customer pay the freight!

While you can call up your local paper this morning and get an ad in this evening, national magazines work months ahead. If you want to cash in on the Christmas trade with an ad in a big national magazine, better contact 'em in June! It is wise to get professional help with writing and placing ads.

NBA COMPUTERIZED CLEARINGHOUSE

Keep in touch with NBA. Read every issue of *BUFFALO!.* While NBA's computerized clearinghouse is developing, this is the best way to keep your finger on the pulse of the market. Marketing will be much easier and more profitable when your Association gets this service up and running.

The computerized clearinghouse is a marketing co-op. Members will subscribe whole carcasses, sides, quarters, or "x" number of pounds by the week, month, or annually. Guidelines to insure quality control are being developed right now. Things like the age of the animal, days on grain, grade, and so forth that will be acceptable to the co-op. Also being developed is a pricing schedule for buffalo meat being sold through the co-op. You won't set the price, the NBA will. It will probably be far higher, however, than the going rate for buffalo meat in buffalo country, as the market targets are the East and West coasts where buffalo meat is presently bringing about double what the average buffalo producer is getting in the midwest by selling out of his freezer to the local market. You will pay a subscription fee for the service—somewhere in the neighborhood of 10¢ a pound. You will also be obligated to pay freight to get the product to the consumer, and you will need to have the meat processed according to NBA specifications.

In return for your trouble, you will realize a guaranteed market at far higher profits than selling out of your freezer to the local market. Additionally, you'll get national advertising benefits. In return for your 10¢ a pound, NBA will provide a computer to store information on who has what, when, and where, plus a staff to keep track of what is going on.

NBA, as a part of this program, will also provide a toll free telephone number so sellers will be able to keep track of daily changes. This free phone number will also encourage consumer questions and purchases.

Besides a guaranteed market, your most tangible benefits from the marketing co-op's service is going to be an organized national advertising campaign highlighting your product, the kind of advertising few individual operations even dream of affording! Even with initial setup costs, the NBA expects to be able to pour more than $100,000 into advertising the first year.

The NBA's initial advertising thrust will be directed toward restaurant trade magazines. From there NBA will branch out to specialty product manufacturers such as Stouffer's with a punchy, colorful brochure; then to national consumer and women's magazines such as *Ladies Home Journal*. The demand is there, but it must be developed; the supply is there, but it must be pooled! The NBA is prepared to do *both*, and expects to have the buffalo meat marketing co-op functioning by early 1984. The computer will also be utilized to store information on buyers and sellers of buffalo by-products, thus insuring subscribers maximum profit from their entire buffalo yield.

ABSENTEE OWNER OPPORTUNITIES

by Bill Moore

We, Bill and Pat Moore of the Bar X Ranch near Cody, Wyoming, started an absentee owner's program as an outgrowth of interest from enthusiastic buffalo buffs.

How does a person become an owner? He or she must have a place to keep this new pride and joy, conversation piece, status symbol, tax shelter, investment, or meat product. Believe me, there are as many reasons for keeping buffalo as people to keep them.

Many people have the money to invest in buffalo, but not a place to keep them. Even if they can afford the land, fences, and feed, there is the need to check on them and occasionally work them—which, for many, is beyond their resources or area of expertise.

If a rancher has the inclination, he can find interested owners to raise buffalo for. Frequently, he'll be assisting people who are starting from scratch and raising their buffalo besides.

"Why should you raise buffalo for someone else?" you ask. I will be the first to admit there are disadvantages and advantages to an absentee owner program. I think it's evident we personally feel the advantages outweigh the disadvantages. The pluses include, but are not limited to: a sharing of the risks, capitalization without interest, better utilization of your land, a broader base of operation and good record keeping techniques. The less desirable aspects are: less control of culling and dispersion procedures and increased paperwork.

So that you may decide for yourself if running an absentee ownership program is for you, I'll go into more detail on these advantages and disadvantages.

The biggest advantage to us has been increasing our herd size to take advantage of our pastures without borrowing money to buy stock. We had a 100-cow operation in 1977 with 11 head of buffalo and 90 head of lease cows. We didn't want to run cattle and buffalo together as we couldn't get a real handle on what grasses the buffalo were using, what kind of diseases may be attributable to the cattle, etc., but we could not go out and buy enough buffalo to bring our ranch to capacity without borrowing a lot of money at high interest. Besides, at that time, our bankers were not too keen on the idea of loaning money for buffalo.

We started advertising by making up a single sheet brochure proposing that we could raise buffalo for the price of a mug of A & W root beer per day. Root beer was 35 cents a mug and we could raise buffalo for $10 a head per month. We are still doing it; though now we get $18 per month as A & W root beer is 60 to 75 cents per mug. I've often wished we had tied our prices to a gallon of gasoline!

We've tried to keep a balance so the absentee owner can expect a 20% average profit, a figure most of our owners exceed. As with all averages, there will be some below as well as some above. As an example: If the ranch experiences a 5% death loss in a given year such as six calves and two cows (5% is about our average), this looks good for the ranch. However, to the owner who has only two cows and loses one of them, or only one cow and loses the calf, the hurt is all in one place.

Even though one of the advantages of having absentee owners is being able to share the risks, it is, nonetheless, a difficult job to notify the owner that his buffalo died or did not calve this year. One owner purchased sister cows that calved for the first time last year and, for no apparent reason, both lost their calves before the year was out. The cows are extremely good looking cows and we will really be watching them closely this year. Obviously, this owner did not do as well as one of our first owners who started with three pregnant cows and two years and two months later, had a buffalo herd of nine head with a high percentage of females.

A broader base of operation and the better utilization that I mentioned earlier can be obtained by being able to have a more certain and

Curly won his position as Bar X herd bull by personal prowess.

Bar X's early advertising efforts for absentee owners.

accountable income. You will know how much money you can expect with a given amount of stock and plan accordingly. This is barring drought, flood, increased gasoline prices, fencing needs, or the 1,001 items that affect you as a rancher.

Record keeping is both an advantage and a disadvantage. I count it as an advantage that we have had to keep more detailed records. Herd administration and management are so much easier as a result, but it does take time. If you do nothing else with your absentee owner program, make sure your records are correct and in enough detail and do everything strictly straight arrow.

We have every buffalo listed on a separate card with its entire history: age, where born, parentage if possible, sex, ear tag number, registration number, owners, when and what she calved, etc. We have photo albums that these cards go into. All Bar X buffalo are in one book, absentee owners in another and separate books are kept for our associate ranches. When we move the buffalo from one location to another, we move the cards. When an animal leaves the herd, dies or is slaughtered, its card is put in a separate file.

I also keep a daily personal journal that I fill out in a few minutes each night with the major events of the day. It lists sales of stock, butchering, hay purchases, etc. Example: "My son and I hauled 5 T of hay from Ken Barcher's today" or "Sunday—pulled through." This record has been a great help to us through the years as it gives information on calving times, when we start feeding in the fall or quit in the spring, dates of working the buffalo, etc. This record is valuable to us, personally, but it is even more important with an absentee owner program as you are more accountable for your time and effort.

Be careful that you are not involved in what can be interpreted by the Securities and Exchange Commission as a security investment. You may want to check with them or your attorney on your program. In general, you must not have over 30 absentee owners and your literature must describe the risks as well as the advantages of your program. In other words, do not promise anything you can't produce.

CONTRACT KNOW-HOW

Probably the most frequently asked-for piece of information is, "What does your contract look like?" Other frequently asked questions are, "How do I collect pasturage? When and what price? How do we keep records?" (See the reproduced sample.)

I maintain a folder for each of our absentee owners. At the present time we have about 23 owners who own from one to 16 buffalo each— therefore, we maintain 23 folders. Each folder contains most of the correspondence that I had with the owner. On the front of the folder is posted all money that I have received from them and what it was for: pasturage, registration, vaccination, transportation, etc., and a chart of months with payments for basic pasturage. (See diagram #1.)

I bill once a year, usually in December, when I also give the owner his annual report on the animals according to ear tag numbers. My rates are $15 per month per head, paid in advance, or $18 per month per head

	1981	1982
Jan	$30⁰⁰	$45⁰⁰
Feb	$30⁰⁰	$45⁰⁰
March	$30⁰⁰	$45⁰⁰

Diagram #1: Payments for pasturage.

received after January 15. In most cases the December bill is the only bill I send, as most pay in advance by the year. On this bill are any other incurred charges such as cow/calf pair makeup, vaccination fees, or back pasturage.

We vaccinate for Bangs (brucellosis) and with "7 Way." This is a trade name vaccine that protects against a combination of seven diseases. Our prime reason for using 7 Way is to protect against clostridium.

Since we've also been frequently asked what our history cards look like, a sample is duplicated. Each animal has its own card with its entire history recorded on the card. Sometimes, I'm the only one who can determine what the history is, but it's all there. The cards are kept in a graduated plastic card file where only the last ½ inch or so of the card shows. At a glance I can see all the cards which show the owner, the sex and year born of the buffalo, the registration or certification number of the animal (if appropriate) and that animal's ear tag number or name. By flipping the preceding cards back, I can see the entire history of an animal since it came to the Bar X.

In the case of the cow's sample card (Marie), we have kept track of her calving record each year and where those calves went. The buyer is indicated in ( ). Also on the card will be the sire of the calf and any other pertinent information that is appropriate. Calves not sold and retained in the herd will then become recorded on a separate card. If the animal dies, that information is recorded on the back of the card and those cards are retained in a separate file.

SAMPLE CARD

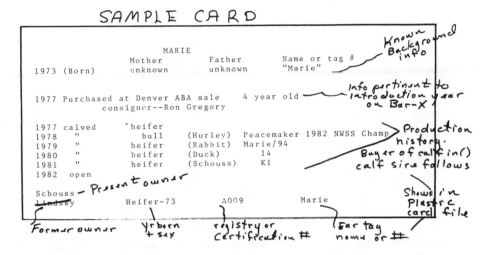

I will be glad to answer any questions of anyone who would like to inquire between 5 a.m. and 7 a.m. (MST). That's the best time to catch me in the house. I'll also answer short questions accompanied by a SASE. Call me at: 307-645-2471 or write Bill Moore, Star Route, Belfry, MT 59008.

AGREEMENT TO RAISE BUFFALO AT THE BAR-X RANCH

This agreement, made and entered into this_____day of_____, 19_____, by and
between_____, hereinafter called owner, and Pat and Bill Moore at the
Bar-X Ranch, hereinafter called Bar-X, WITNESSETH·
The purpose and intent of this agreement is to provide for the delivery to the Bar-X certain
buffalo for their management and care.   Now therefore, in consideration of the mutual
promises and covenants to be performed by the respective parties, BE IT AGREED:
1. The owner will deliver to the Bar-X at its property at Clark, Wyoming the aforesaid
buffalo in the number of_____, consisting of_____.
2. Upon delivery of said buffalo, the Bar-X shall maintain range, feed and care for them
in accordance with the customary and proper practices of buffalo management.
3. The Bar-X is to receive for said services $_____per animal per month or $_____ per
month for cow/calf pair. When calf is five (5) months old it will be considered as if mature
and will be billed at the above animal rate. All payments will be collected in advance and
any animal whose pasturage fee is six (6) months in arrears may be sold after serving due
notice to owner. The proceeds of said sale will go first to pay incurred pasturage and
sale costs with the remaining monies to be sent to owner. The Bar-X retains the right to
change the pasturage fee as costs may dictate.
4. All buffalo will be tested for brucellosis and be brucellosis free and in good health
upon arrival at the Bar-X. Animals that come in from out of state will then have to be
tested again no sooner than 30 days nor later than 90 days in order to meet state regulations.
The expenses involved with such testing will be borne by the owner.
5. Bar-X agrees that it will carry liability insurance at its expense and in sufficient
amount to protect the owner from liability that may arise by reason of acts of the public
generally, third persons and trespassers.
6. The owner is to be responsible for all taxation levied against the animals. Insurance
for death loss or other perils of the animals is also the responsibility of the owner.
7. It is hereby agreed that it is the principle intention of the parties that such buffalo
shall be maintained, cared for and handled as long as practical in accordance with the normal
and customary practices of such an operation. The owner hereby agrees that the Bar-X shall
not be held responsible or liable for losses of such animals occasioned by accident, unlaw-
ful acts of third persons, natural causes or acts of God.
8. It is specifically understood and agreed that Bar-X shall have at all times the exclusive preference option and right to purchase or refuse to purchase, as the case may be any
and all animals offered for sale hereunder at fair market value.
9. The Bar-X will make an annual report of the animals to their respective owners annually.
10. The Bar-X will assist owners to brand or tag their animals, aid, or obtain veterinary
aid, test, or otherwise handle the animals, assist in marketing or butchering, provide
pictures or help register the animals at the owners request and at the owners expense.
11. It is mutually agreed that in the event that any or all of these animals should become
wild, fence jumpers, or mean to the point of being a hazard to the Bar-X or to the general
public, the owner will be notified to regain possession of such animals or to order the
disposition of said animals that are the cause of hazards mentioned above.
12. The parties hereto agree that they will fully cooperate in every means practical and
possible to the end of upgrading and promoting the improvement of their buffalo herds and to
propagate the increase thereof.

IN WITNESS HEREOF, the parties have hereunto set their hands on the day and year above
written.

_____   _____
OWNER   BAR-X

194

17
Capturing More $$ With By-Products

Working taxidermists have kindly shared their experience with us to help get the finest trophy head, robe, or leather possible. Note that here, as in all buffalo matters, opinions differ (i.e., one wants frozen heads, another doesn't). To be safe, follow the instructions of your own taxidermist. Values and prices given will doubtless be quickly obsolete. The trend is nowhere but up! Checking around will bring you surprises, their degree of pleasantness depending on whether you are selling or buying.

CAPTURING HEAD & HIDE PROFITS
by Dave Raynolds

How many buffalo ranchers are missing those important profits from heads and hides that established ranchers regularly collect? Discussion with a number of relatively new NBA members suggests that this route to profits is often overlooked and usually misunderstood.

The retail value of head and hide is nearly as great as the retail value of the bufflao's meat! Although the tanning and mounting procedures required to reach these retail prices go beyond the effort the usual rancher wants to invest, he can—by cooperating with the nearby taxidermists—reap some of these profits without unduly complicating his operation.

The key to pricing buffalo heads and hides is their comparative scarcity. Probably fewer than 2,000 heads, and an equal number of hides, reach the taxidermy stage each year—that's 10 of each for a million Americans! Buffalo ranchers are in an excellent position to capitalize on this scarcity. Most taxidermists have never done a buffalo head. Of those that have, the buffalo head probably represented the highest priced mounting in their shops.

Taxidermy involves a number of steps and about two days of work for a head and hide. Ideally, your taxidermist should be on hand to handle the caping, which consists of cutting off the hide just behind the shoulders

195

and taking it and the head as a package back to his shop. The cape is the heavy fur from behind the buffalo's ears to the middle of the front legs. Taken at its prime, it makes the best robes and rugs. A robe is worth four times as much as leather. (The rancher should remember to get the tongue and brains at this stage, or sell them to the taxidermist.)

The remaining hide is usually removed with a belly slit, although it can be done Indian style, which was to make the main incision along the backbone. The tail should be left on the hide when skinned whiteman style. The feet can be utilized separately, either by the rancher or the taxidermist.

Back in his shop, the taxidermist takes careful measurements and orders head form, ear liners, eyes. He skins off the head skin and cape and sends them to the tanner. Forms may be made of papier-mache, fiberglass, or styrofoam. These materials and tanning cost the taxidermist (in Wyoming in 1980) about $125. The rancher should be able to net $600 on the sale of the mounted head and tanned robe. Or he can have the taxidermist sell them for him.

It is an advantage to both the rancher and the taxidermist to book a year's work well ahead, since buffalo harvest usually comes during hunting season, when the taxidermist is his very busiest.

The old trophy bull, although his meat is less succulent and therefore lower priced, yields more of it and should be worth more dollars. The head and hide returns should be much higher. The extreme case would be a full mount of a head eligible for Boone & Crockett entry. Hide returns should be larger because there are more square feet. Heads should be priced individually because fewer than one trophy head per million Americans comes on the market each year.

When a buffalo is harvested outside the prime season, the head probably will not make a good mount. An alternate marketing plan is to boil the horn casings off the skull. The bleached skull should bring $100 or more. The horns can be marketed separately as powder horns for muzzle loaders. The hide, tanned with hair off, still has value, although less than the winter robe. There is a special, though limited, market for these hides among Plains Indians for authentic tipi building; a full tipi requires about 21 hides. (My thanks for description of procedures to Wayne Robillard, Wind River Taxidermy, 824 Main St., Lander, WY 82520.)

CHOOSING & TREATING THE TROPHY HEAD
by Thomas J. Matuska, Taxidermist

A trophy may be a bull head or a cow head, hooves, horns, robes, or some other item. Trophies, like beauty, lie in the eye of the beholder.

Trophies take the form of magnificent head mounts, half mounts, and full mounts, along with novelties such as buffalo robes, rugs, blankets, leather products, leg lamps, and even footstools. The list is limited only by your imagination, ranging in cost from a few dollars to a few thousand. By far, the most popular and most frequently done by our studio is the head or full shoulder mount.

Considerations for Quality

PRIMENESS: A prime coat is taken when the buffalo's woolly hair is at peak in length, color, and texture, usually during the bitterest cold. Most buffalo in the Midwest reach their prime in January and February, just prior to shedding.

BEARD: Bunk feeding tends to wear the beard short—important to some.

Caring for the Trophy Head

Ideally, your taxidermist receives a bison head intact, with the skin on the skull, the head simply severed from the neck. The shoulders may be skinned out to the base of the skull, to prevent any waste of meat and to give the taxidermist plenty of length to work with.

If there will be a delay in getting the head to your taxidermist, have your local locker plant freeze it. Wrap the frozen head in plastic and deliver as soon as arrangements can be made.

If your taxidermist is too far away for you to drive, you may want to skin the entire cape from the skull, to minimize the freight bill, and ship it. Give special attention to the ears, lips, and nose. Salt the cape immediately to prevent any spoilage or hair slip. This procedure requires great care; don't try it until you've consulted a professional.

Choosing Your Taxidermist

Far more important than anything we've talked about so far is how to choose the person to handle your trophy, for no matter how much time and effort you spend choosing it, or the great care you lavish on skinning and delivering, it will all be in vain in the wrong person has the responsibility.

Taxidermists are a dime a dozen, but experienced master craftsmen—professionals—are one in a thousand. Don't fear to ask to see his credentials. Was he properly educated? Does he know his business? Does he have a broad knowledge of wildlife in general and buffalo in particular? Just because he's called himself a taxidermist for 30 years doesn't qualify him for your trophy; 30 years of poor workmanship and low standards make him no better than the novice.

Does he make his living at taxidermy? If not, find out why. Stay away from hobby or part time taxidermists. They often lack the expertise to care for your trophy properly.

Does he keep his work moving? Taxidermy demands hours of intricate detail, and must not be rushed. On the other hand, you can't be expected to wait two or three years for your mount. Check his reputation for speed, efficiency, and consistently good work.

Stay away from bargains. Taxidermy work is not cheap, nor is that 2,000-pound bull. Ofttimes, cheap or "bargain" mounts may look good on the outside but have inferior materials inside. This can result in a good looking head that may fall right off your wall . . . one piece at a time! As in most things, you get what you pay for. Remember:

> The bitterness of low quality
> remains long after
> the sweetness of low price
> is forgotten!

A trophy, artistically done, anatomically correct, that has had the most modern materials and techniques incorporated into it, is worth many times the cost of mounting—a fine investment.

With minimum care and an occasional cleaning, a modern mount will easily outlast its owner and his grandchildren and remain a constant reminder of a magnificent beast for years to come.

For your free booklet on skinning, salting, and shipping your head and hides, write to: Matuska Taxidermy, Center Lake Drive, Spirit Lake, IA 51360.

CARE AND VALUE OF HIDES & HEADS

by Chuck Lietzau, Taxidermist

Experience in buying buffalo hides, heads, and skulls for taxidermy and resale over the years has convinced me that the buffalo raiser is in a unique position. He has a potential steady profit year after year from the hides and heads of the animals he harvests.

Consider that the raiser of domestic stock, selling to the livestock markets, has no control over the extra products his animals produce. The buffalo raiser can turn them into nearly clear profit with little extra effort. I hope you will keep the following information and refer to it when considering the best way to get the most from the animal that is truly a part of our American heritage.

Heads

The use of the buffalo head as a trophy is based on knowledge acquired over 17 years. Although my business is custom taxidermy, the opportunity to sell a number of heads taught me that it can be a profitable addition to our services. The larger head, from a well developed three-year-old bull, or the older herd bull, is the head most buyers are interested in. The smaller bull head may be sold for a nice profit also, but as people become aware of the difference between the two-year-old and a four- or five-year-old, it will be the larger head that is desired. Cow heads for mounts are not desirable in most areas.

The bull can be butchered and the head sold for mounting. Or you can sell the complete animal to someone who comes to your ranch to shoot the animal for a set fee covering meat, head, and hide. In either case, the animal should be properly skinned and the skin and head cared for.

Skinning the Head for Mounting

I prefer receiving a buffalo head for mounting as follows:

I like it skinned as shown in diagram #1, making a cut completely encircling the body just behind the front legs. A second cut down the back of the neck begins just behind the skull and joins the first cut at the top of the shoulders. These cuts should be made before the skinning is started; from this point, proceed as usual.

Diagram #1.

When removing the skin from the rear forward, the rear half will separate from the front when the cut behind the front legs is reached. The cut on the back of the neck will open as the skinning is completed, making it easy to reach the back of the head and sever it from the rest of the carcass.

You will now have two pieces: the rump hide or back half, and the head and shoulder hide together, which is called the "cape." The rump

199

hide should be salted as described below. It is best to leave the head unskinned and simply salt the neck and shoulder hide attached.

Since most trophy bulls are harvested during cold weather, it is ideal to bring the head to the taxidermist within a day or two, before it freezes solid.

Values

The raw head and hide handled in this manner should be worth between $125 and $250, perhaps more. The old bull has the most potential for extra profit: From $50 to $90 for the prime hide, and for a good skull. The animal taken outside the prime season can be expected to yield a hide about half that.

Hides

Dealing in fairly large numbers of hides over the years has taught me a number of things of value to the buffalo rancher. The hide is the most marketable part. To get most from a hide, the animal should be killed between December 1 and January 10, when the hair is at its best and the long hair on the shoulders has not yet begun to loosen. Expensive experience taught me that, when a hide is taken after January 15, enough of the long hair on the shoulder will be lost to reduce the value.

Let's say, for now, we're considering killing cows and/or prime butcher animals about 30 months old. If you are planning to sell the hides for hair-on (robe) tanning, you cannot use the head for a full shoulder mount; to make such a mount, the cape will have to be taken, leaving only the rear half of ⅔ of the skin which has little value for hair-on tanning and is worth only about $10 for leather.

Skinning

This is very important, as a hide damaged with excessive holes may be worthless. Leave the tail on. Cut the head off directly behind the ears. The heavy fatty portion at the base of the tail and the tail bone should be removed.

Preserving the Hide

Unless the tanner is going to pick up the hides within a couple of days, they should be salted with fine table-grade salt. Hides should always be salted in the shade, out of direct sunlight.

Lay the hide hair-side down with the center higher than the edges so the liquid the salt draws out can run off. You can mound up a small pile of hay or straw to lay the hide on. Or lay a 6-inch post about 8 feet long on the floor of the salting building and nail two old sheets of plywood to it, forming a "roof" over which you lay the hides to drain.

Salt the entire head, reaching the very edges, as they will often want to fold over or stick together, thus escaping the salt and spoiling. Don't forget the tail. The salt kills bacteria that would cause the hair to slip. Better use too much salt than not enough: 20–25 pounds. Hides salted and stacked in this manner, so they will drain, can be stacked with no danger to the hide, but should be sold or tanned before warm weather.

Leather

Hides unsuitable for hair-on tanning should be salted as above and can be tanned for leather.

Shipping Hides

When you must ship hides to buyer or tanner, salt as above, then leave them for about three days. Next, hang the hide over the rail (in the shade) and let the excess moisture drain and evaporate. Hang for a couple of weeks if necessary, but take it down and fold it or roll it tightly before it becomes too dry and stiff.

To keep from getting salt all over your new pickup, put the hide in a plastic bag and then into a burlap sack and ship. Hides dried as above can be stored for months EXCEPT they should not be stored in plastic, but in large cardboard boxes away from dampness. If storing over summer, a can of moth crystals sprinkled into the box (providing the cover is tight) will prevent insect damage.

Skulls

The buffalo skull is also of value. To avoid making holes in the forehead, shoot the animal from the side: between the eye and the horn, and 2–4 inches behind the base of the horn. This shot placement is usually more effective than the head shot.

Values

The buffalo market is so bullish that prices given here will doubtless be much exceeded in the near future. As of this writing the retail value of a prime tanned buffalo hide can run from $400 to $800 and the value of a prime old bull in a full shoulder mount may be $1,000 or more.

I will deal, however, with the more realistic prices that you, the rancher, can derive from a larger number of animals with minimum effort. You can seek the higher retail prices, but this involves laying out more capital and marketing the finished product. But, remember, any profit made is better than *giving away* the extras, as you did on the last load of beef you sold.

A prime cow hide should bring $50 and up, the 30-month-old bull up to $75. Animals kept in close confinement and showing rubs or excess manure in the hair will suffer in value, especially if the hide is to be tanned hair-on.

Diagram #2.

A hide not suitable for hair-on tanning should be salted in the same way as above and then can be leather tanned. The price for a leather quality hide would be about $25. My knowledge of hides harvested in midsummer is limited, but I feel that the hide would be quite thin and of lesser value. There is a definite demand for certain types of buffalo leather. To obtain the quantity of raw skins to produce the different leathers, and to find a reliable tanner, are problems.

Skulls should bring $15 to $25 for a two- or three-year-old bull. Cow skulls are worth about half that.

The use of bulls four and older has two advantages: 1) The hide can be tanned hair-on for a robe, but you must leave the hide full size and remove the head. This larger hide will be worth a minimum of $75. 2) The skull in good condition should bring $25 to $35. If of trophy proportions, it will bring top dollar. Thus, the old bull has the most potential for extra profit.

Odds & Ends

Other parts that may have some value are the feet, some of the bones, and the large tendons. Two of these tendons lie along either side of the backbone and also on the back lower part of each leg. These tendons are used to make sinew, the thread used for sewing by the Indians. This would now be used to recreate their authentic garments.

We have encouraged buffalo raisers to make feet available at no cost to explore the possibilities for their uses. I would also like to do this for sinew; it could perhaps open a market.

For more information, inquire of Chuck Lietzau Taxidermy, Cosmos, MN 56228.

SELL BITS & PIECES

The Indians used every scrap of buffalo. You can sell virtually every part at a healthy profit. Black powder shooters treasure the horn sheaths for powderhorns; scrotums for bullet pouches. Tails make switches; hooves make gunracks, bookends, and lampstands. Scrimshaw artists pay for toenails. Even the wool hanging on brush and fence barbs has its market! The bull has a yard long cartilaginous core to his penis which, when dried, makes an interesting, effective and salable whip! The Indians even used stomachs and bladders for pouches and bags.

Interestingly, Lloyd Wonderlich has a "standing order" from a buckskinner for all the buffalo toenails he can boil off the hooves! They are used by scrimshaw artists as the sailors in the days of sail used walrus tusk.

NBA will help you locate manufacturers who will turn your buffalo leather into salable products! It is important to find a reliable tanner. Service from tanneries ranges from two to 12 months, from reasonable to expensive, and from excellent to shoddy workmanship. Take time to do some checking and comparing before you ship your valuable hides.

To maximize your hide profits, it is best not to brand. The brand is burned scar tissue and hides tend to break open there. The damaged area must be cut out and hand sewn—very costly! If you must brand, freeze branding causes less damage. Also beware that the hair comes off of bruised areas . . . even the bruises where the animal fell when shot.

BUFFALO WOOL PROVIDES NEW INDUSTRY

by Mort Hamilton

Historically, the buffalo provided everything the Indian and pioneer needed—food and shelter. Hides were used for blankets, teepee walls, and rugs. Occasionally, the outer hairs were twisted into rope, but the Plains Indians were not weavers. Strangely enough, the wool and hair were never separated from the skin and woven into useful articles.

The commercial weaving of buffalo wool and hair was a non-existent craft until 1979 when Grete Heikes, a master weaver and native of Norway, moved to Fort Pierre, SD. Much of Fort Pierre's 18th Century history is based on the great American bison. Her weaving studio now sits on property that was once owned by Scotty Philip, credited as one of the saviors of the buffalo. The world famous Triple U buffalo herd owned by the Houck family roams the prairie near her home.

A diversified operation, the Houck ranch has been innovative in many aspects of buffalo promotion. Their processing plant, built in the early 70's, now ships meat and hides throughout the United States. Even in their home, the Houcks have utilized buffalo hides in the reupholstery of their antique furniture.

Always alert for new fiber material, Ms. Heikes began research on buffalo hair and wool. Her background and preparation for this work actually began in 1964 in Norway and continued in Denmark and England under the master/apprentice craftsmen system.

While visiting the Houck ranch, Ms. Heikes and Ms. Lila Houck found they could grasp handfuls of the wool and hair as the animals moved through the chutes at the Houck processing plant. A small bag of wool and hair was gathered this way and the new project was launched!

Preparing the fiber for weaving involves a laborious process of hand-separating the inner wool from the deck hair. The wiry outer hairs (about the quality of fishing line) were not suitable for this first project. Ms. Heikes found that the sack yielded about ⅓ wool and ⅔ hair and debris. The wool was discovered to be very lightweight, soft, and beautifully colored. Hand spun into 1-ply yarn, the wool was then washed, made into skeins and set up for the weaving process.

In the first efforts, Ms. Heikes combined the lovely brown buffalo wool with woolen yarns into warm, soft scarves and shawls. Through experience, Ms. Heikes found the innerbody wool to be most luxurious and most suitable for clothing worn close to the skin. The coarse outer hair has been spun and will be woven into saddle blankets or heavy outer clothing. A wall hanging made of buffalo wool and white sheep wool was produced in the overshot technique. Then she wove a shirt-jacket of sheep and buffalo wool decorated with unspun tufts of wool.

Ms. Heikes has applied for a grant from the National Endowment

Grete Heikes carding buffalo wool.   Grete spinning the carded wool.

for the Arts for her buffalo project which she plans to develop into a full line of clothing. She will maintain a full time studio with apprentices to continue her work.

Accessibility of the raw material is, of course, very important and the Houck Buffalo Ranch is cooperating with the weaver's efforts. The extensive Houck herd is contained to the point that the wool may be gathered as it falls here and there over thousands of acres when the buffalo shed, or can be picked off through the chutes.

Recently, the Houcks called Ms. Heikes to ask if she was interested in shearing a dead buffalo. Interested? You bet! Regular sheep shearing equipment was necessary and the chore yielded one very large bag of wool, hair, and mud. Only partial shearing was possible due to mud and debris. Nobody has yet volunteered to pull the hair off a live, standing buffalo.

The Houck family is quite excited about this new phase of the buffalo industry. The animal will now truly provide more than primitive man ever dreamed of—a new sophisticated line of clothing and outer wear.

18
Hunting:
An Easy Buck

One profitable way of marketing your big bulls is to sell hunting privileges. While some consider this method "inhumane," ranchers find it's kinder to the big bull to shoot him where he stands than to try to round him up, push him through the chute, kicking and fighting all the way, and into a truck where he may break a leg. And it's sure a lot more humane to the help who would otherwise run the risk of cuts, fractures, bruises, sprains, gores, concussions, and hemiplegia trying to bring the monster to the block.

One big game hunter, who had shot lions, tigers, elephants, rhinos, and polar bears, told me that walking out into the feedlot and dropping a standing buffalo was the biggest thrill of his life. He didn't count the cost, either. "The expensive part is that little $10,000 diamond trinket the wife expects when you get home."

Methods for hunting range from walking into the feedlot to scouring the mountains on horseback, camping out, and packing the spoils home by mule. You'll find takers no matter how you slice it.

Here's how some NBA members work out the details:

HOW DURHAM RANCH HANDLES HUNTERS

by Jack Errington
Assistant General Manager
Durham Meat Company

Each year we cut out 10 to 20 big bulls with nice big heads and horns and schedule hunting for November, December, and January, when the pelage is prime. We schedule the hunts through a commercial guide.

We furnish all vehicles and equipment necessary to complete the hunt. For those who want a head mount only—we field dress after the animal is shot, haul into our dressing facilities and remove the hide and head. The carcass is then placed in the ranch freezer for later shipment to California for processing into ground meat. Our current fee is $1,500 to $3,500, according to the size of the bull and the horns. For the higher fee, the hunter gets to keep the meat, too.

We have hunted by horses, but we always send vehicles along as a safety precaution. This pays off from time to time. We don't carry special insurance. Each hunter signs a release and holds us harmless. We have an umbrella liability policy that covers hunting.

SUB-ZERO PRAIRIE BISON – vs – .44 MAGS
by Albert L. Pfitzmayer

Several years ago, if someone told me of the high performance special SSK penetrating round KTW (modified) I would have been puzzled. Thanks to J. D. Jones, I was furnished with six of these unbelievable rounds.

I had previously mentioned to J. D. that I was about to embark on my eighth African Safari, and shooting exhibition, performed for the enjoyment of the East African tribesmen at their government's request. I always bring along my 29 S & W .44 Mag. as a backup on mbogo's (Cape buffalo) or whatever heavy stuff that might be encountered.

On prior East African hunts I had the good fortune to collect all of the big five (rhino, Cape buffalo, lion, elephant and leopard). However, the .44 Mag. seldom came into play, save only a few tight moments when the situation was close-in shooting. I guess old J. D. knew better than I, as shortly after my mention of the African upcoming hunt, he sent, in a cute package, six green tipped cartridges along with a letter of what to do should a rogue bull elephant appear tent side. Along with my usual armament, the six Special SSK-KTW rounds fitted in my speed loader joined the hunt.

Two record class masai lions, plus a very respectable Cape buffalo, were destined to fall to three one-shot kills, all dispatched with J. D.'s knowledge and trust in a bullet that is in a class all by itself. Since that time, I cherished the remaining three rounds of KTW's. Subsequent East African safaris presented little opportunity to use the .44 Mag.

In 1981, I was notified by my lifelong hunting partner, Kenny Gerstung, of Mequon, WI, that he had booked a bison hunt out in frigid South Dakota. Naturally it would be my pleasure to join Ken again. As always, the .44 S & W would go along. We were advised by Mr. J. Houck of the Triple U Ranch that they would reserve the right to the second shot, should our fingers be too cold, guns not sighted in, or whatever the usual excuse be. Four of us departed from NY with the heaviest artillery we could muster: two 300 Weatherbys, one seven Mag., and a Colt Sauer 458. Oh yes, and three S & W revolvers, of .44 Magnum bore.

We arrived at the Triple U Ranch, along with an Arctic snow storm. The wind blasts brought the thermometer down to –20°. We were not eager to get out of our cozy rented car. Mr. Houck greeted us, and after a brief exchange of bison frequents, we were off. Riding in back of an open pickup truck is one way to quickly acquire frost bite, so two of us elected to hunch in back of the cab, while our guide drove the pickup and tried to locate our reason for the hunt.

Ray Maneuso neatly picked off a magnificent bull standing on a ridge at about 250 yards. We made our way cautiously up to the spot where the now almost frozen bull rested. On the way up, I asked Mr. Houck if he had ever dispatched a bison bull with a heavy caliber handgun? His near frozen eyebrow raised and he looked down at the snow, stammered a bit and replied "Nope!" I fished into my DeSantis shoulder rig and eagerly produced my 629, the presentation given me by Smith & Wesson as one of the top 10 in the 1980 Outstanding American Handgunner awards.

Mr. Houck inspected the new shining 629 with keen interest. We held our breaths as he replied, "Can ya make a 60-yard shot in the ear with that pretty thing?" I replied that I thought I could. He stated, "You can see by the movements of the bison, they are pretty spooky. Ok, I'll try to get you in on the far side of a draw, then you will have to sneak up on 'em."

It was Kenny's turn to shoot. I fumbled around in my snow filled pockets and came up with one SSK-KTW Special. Fancy shooting Kenny caught his big bull with three cows moving along an ice covered stream bed. The bull started to gallop up the draw and Ken let one go with his 29, smacking the bull a shade low. The SSK-KTW broke both jaws, ricocheted down and came to rest in the bison's chest. Kenny's next round, a handloaded Barnes FMJ 300 grain solid slammed into his neck and he dropped abruptly into the drifting snow. Mr. Houck seemed very impressed. Kenny cursed at having to use two rounds.

After a full inspection of Kenny's trophy of a lifetime, we were off again. Ed Vanderhyde of police combat shoot fame was also given one SSK-KTW for his 29, and after two long, frigid hours we came upon another small group of standing bulls, E.Q. elected to take the biggest bull in the herd and after stalking to about 65 yards, his .44 let go. The SSK-KTW drove smack into the bull's ear and he dropped into the snow, stone dead. He came to rest on all fours, hesitated and slowly rolled onto his side, finished.

It was now my turn to complete the group of excellent one shotters, not to mention Kenny. I could see it wasn't going to be easy. The wind had picked up and the day was about to close. Traces of darkness haunted the draws and after another miserably cold hour passed, our guide decided to turn back as we had a long way to go before we reached the main ranch house.

On the way back we stopped on a high ridge and glassed the low ridges in the area where Ray had taken the first bison. Out in the distance, the bison appeared to be tiny specks of black pepper, but as we closed the range there could be no mistaking one of the largest bull bison of the Triple U Ranch. I dropped off the back of the pickup and hunched up in the most inconspicuous way I could. I guess the Sioux would have been hysterical watching me, but I was so damn cold I just couldn't get down any lower or move any different.

Luck was with me; the bull we spotted was probably the old herd master, and it was his duty to investigate whatever was trying to crawl up on his herd of cows. The last of the SSK-KTWs went on its way, meeting the bull behind his left ear, and for all I know is now circulating over Peking, China, because a buffalo head won't stop them. Anyway, the last of four magnificent bull bisons was now a legend.

All three SSK-KTW rounds had penetrated over 16 inches of very formidable bison resistance, truly a bullet of special performance!

Ed Vanderhyde, Ken Gerstun, Jerry Houck and Al Pfitz-mayer admire Al's 1980 trophy. Their buffalo were taken with STW model 29.44 magnum revolvers.

Editor's Note: The Standard KTW is now manufactured and distributed by Sage International, Ltd., 2271 Star Court, Auburn Heights, MI 48057, 313-852-8733. This bullet is turned from architectural bronze and Teflon coated. It is designed as a metal penetrator for police use and is available only as loaded ammunition in a variety of calibers—including 25 ACP. The SSK-KTW special round is highly modified and, from limited testing, apparently almost doubles penetration in some media. It penetrated 22 inches of elephant skull from a 5-inch barreled .44 Mag. and has killed Cape buffalo with shoulder shots. SSK will not disclose the exact modification of the bullet, but it is very expensive. Each bullet costs in excess of $5.00.

UTAH HUNT: ONE OF A KIND

The Salt Lake Tribune

Utah buffalo hunting is a one-of-a-kind happening. It is the only uncontrolled, unguided, you're on your own, find 'em if you can, bona fide hunt for North America's biggest game animal in the United States.

Oh, there are other buffalo "hunts" on private ranches and in controlled areas where the guide says "shoot that one," but there are none others like Utah where the herd roams free and wild and the hunter is entirely on his own.

Permit hunters can take any buffalo they can find—bull, cow, or calf—provided they can find them. It isn't always easy, but buffalo hunting in Utah the past several years has had very high success.

Utah's first buffalo hunt was held in 1950, several years after a small number of animals had been introduced from Wyoming. No hunts were held between 1951 and 1959 while the herd was permitted to grow and expand, but they have been held annually since then with excellent results.

Until 1978, buffalo hunting was limited to Utah residents only. Nonresident permits have been issued for a fee of $1,000 each. Resident permits cost $100.

Despite the high success rate of Utah buffalo hunting (it has been 100% many times) the hunt itself is often no piece of cake. The herd, now estimated to be nearly 200 animals, roams the Henry Mountains area in Wayne and Garfield counties. It is rough country, some of it accessible only to four-wheel-drive vehicles and horses. The terrain is steep, rising from barren desert at 4,000 feet to heavily wooded slopes at 11,000 feet.

The buffalo range mostly from 7,000 to 10,000 feet from early spring until late fall. Despite their size, they have adapted very well to the steep, rocky and brushy canyons which hide them very well. And, to the surprise of some hunters, buffalo are also very quick and agile and can move in and out of seemingly inaccessible areas with great ease.

While most hunters find their buffalo within the first few days of the hunt, many have reported following buffalo tracks for almost the entire hunt only to find the animals repeatedly located in steep canyons from where they would have been impossible to remove.

Said one unsuccessful holder of a recent permit: "Every time we spotted a bunch they'd head down some steep canyon. We didn't dare shoot one down there because it would have taken a week to get it out."

Nevertheless, buffalo hunting has had an extremely high success rate of nearly 90% over the years. The chances of getting a buffalo are a lot better than getting a permit. Successful applicants are required to take an orientation course from Division of Wildlife Resources personnel prior to the hunt, but otherwise are not assisted in any way.

The biggest problem facing buffalo hunters is how to handle and transport the animal after it is killed. Mature bulls normally weigh between 1,500 and 2,000 pounds, and that's not something you can just drag back to the truck.

Hunting parties normally consist of three or four people, even when only one permit is at stake. It takes that many helpers to skin, cut and transport the meat and hide. A good heavy-duty vehicle is also suggested, with a second one for hauling gear.

Instructions for applying for a buffalo permit are usually announced every year in June, with an application deadline of late July. A public drawing is then held to determine successful applicants. If you want to hunt buffalo in Utah next year, now is the time to start making plans.

For More Information . . .

Annual organized hunting programs are available through the Utah Division of Wildlife Resources. For the most difficult hunt available, contact: Jim Bates, Utah Division of Wildlife Resources, P. O. Box 840, Price, UT 84501. Custer State Park has the largest trophies available. Contact Warren Jackson, Custer State Park, Star Rt. 3, Box 70, Custer, SD 57730. Triple U Enterprises offers the most options. Your contact there is Jerry Houck, Triple U Enterprises, Rt. 2, Box 42, Ft. Pierre, SD 57532. Durham Ranch also offers several options. Contact Armando Flocchini at Durham Meat Co., P. O. Box 26158, San Jose, CA 95159.

Many individual ranchers in the U.S. and Canada also offer hunting on a year to year basis. Frequently, private hunts offer many options and spectacular trophies. For information on hunts available write the NBA, or better yet, place a small classified ad in *BUFFALO!* magazine (it'll cost $5 to $10 to reach 500+ buffalo ranchers). Let them come to you!

RECORDING REMARKABLE ANIMALS
by Dave Raynolds

How big is big?

A common problem for all who observe live buffalo is the difficulty of estimating size and weight. It's not hard to pick out the largest and smallest animals in a herd. However, comparing your recollection of buffalo in one herd (that you may have seen months ago) with other buffalo you're looking at now is much more difficult. As an illustration of this problem, the NBA members tried to guess the weight of a bull at the Catalina Island spring convention in 1981, estimates ranged as much as 40% from the actual weight recorded on the scales.

For many years North American big game hunters have used horn measurements of harvested animals as one way of comparison. Several organizations have established measurement systems for their members. The Boone and Crockett Club bison measurement system is one example. Their first book of measurements appeared in 1932. It represented a substantial improvement over the single measurement records previously published by the London taxidermy firm of Rowland Ward.

NBA members who have discussed horn measurements mention the following problems in the Boone and Crockett measuring system for bison:

—the system tends to include only male animals;

—the high premium on symmetry affects scoring by subtracting "differences." For example, if one horn is an inch longer than the other, both lengths are added together, but then the one-inch "difference" is subtracted from the total score;

—only a few large harvested buffalo have their scores printed (though more are measured). Records show only those taken under special hunting conditions by someone who later got the head officially measured and scored by a registered Boone and Crockett measurer. Hence, the published lists are far from comprehensive;

—horn measurements alone are only one indication of animal size. Weights are also an excellent indication.

NBA has now approved a trophy registration system which will correct these objections and add to the scientific information available regarding remarkable animals.

The adjacent table illustrates how a big bull or cow could be measured. A steel tape should be used, and the horns should have dried for two months or more. Most likely the taxidermist would have boiled off the horn tips in preparation of the mount; this does not affect their size. They should be measured when reinstalled on the horn cores.

While Boone and Crockett records the first two measurements, they are not included in final scores. It is proposed that NBA add in these important measurements. It is also proposed the NBA not use the "differences" that Boone and Crockett assess for lack of symmetry, since buffalo raisers are interested in measuring and scoring animals exactly as they are. These changes in method give the NBA system broader scope, but at the same time Boone and Crockett scores can be converted to NBA scores, and vice versa. This will permit users of the NBA system to compare current animals with historic ones.

SKULL MEASURING TABLE

	NBA Score	Boone & Crockett Score	
1. Tip to Tip	27	27	)
			) not
2. Greatest Spread	35 3/8	35 3/8	) counted
3. Length of Horn			
right	21 2/8	21 2/8	−2
left	23 2/8	23 2/8	
C4. Circumference			
base right	16	16	
base left	15	15	−1
C5. Circumference			
1st '¼' right	13 4/8	13 4/8	
1st '¼' left	13	13	−4/8
C6. Circumference			
2nd '¼' right	11 4/8	11 4/8	
2nd '¼' left	11	11	−4/8
C7. Circumference			
3rd '¼' right	8 2/8	8 2/8	
3rd '¼' left	8	8	−2/8
ADD	203 1/8	140 6/8	−4 2/8
Final score	203 1/8*	136 4/8*	

*This record bull was killed in 1925 in Yellowstone National Park by the chief park ranger.

RECORDING OTHER DATA

The Trophy Buffalo (Bison) System lists other data that could be recorded to ensure a more complete record. In many cases only a few of these items would be available. However, as material accumulated at NBA headquarters, it would be gradually possible to answer questions such as: Who harvests the largest animals in Pennsylvania? Which ranches seem to be harvesting very old animals? Are weight ratios and yields the same for all ranches, or do they vary greatly?

NBA headquarters can probably develop a register at minimal cost—say ten dollars per animal registered—that would be available to visitors at NBA headquarters and would furnish material for comparative articles in *BUFFALO!* Duplicated and attested copies for register entries could be prepared for framing at additional cost; these might be of interest not only to herd owners, but also to trophy hunters.

Once enough large animals have been registered, NBA might take the additional step of preparing a summarizing booklet for distribution to prospective hunters and taxidermists. Ranches accommodating hunters might find such a booklet a useful source of advertising.

TROPHY BUFFALO (BISON) REGISTRY

TABLE II ADDITIONAL DATA

1. Hunter name _____ Address _____

2. Location of hunt _____

3. Owner (private or refuge) name _____
   Address _____
   City _____ State _____ Zip _____

4. Date of kill _____

5. Date scored _____

6. Method of kill (hand gun, rifle, bow, black power) Include specifics: _____

7. Age of buffalo to nearest month or year* _____

8. Sex of buffalo _____

9. Live weight in field before harvest* _____

10. Dead weight in field at harvest* _____

11. Dressed weight before skinning* _____

12. Hanging weight before cutting* _____

13. Skin size in square feet* (will vary depending on caping method-please describe method used) _____

*Optional data, please provide when available.

The Trophy Buffalo Registry reserves the right to reject any or all applications submitted. If rejected you will be notified by mail and your application fee will be refunded.

The Trophy Buffalo Registry will be the sole judge on determining the classification of trophy registrations submitted.

The Trophy Buffalo Registry reserves the right to limit the number of recorded entries per classification.

Recorded entries in any given year are subject to 'bumping' in future years.

RECORDING FEE: (submit with application)
Certificate of hunt data (suitable for framing): $10.00 (mandatory)
TBR Plaque engraved with hunt data: $10.00 (optional)
  $75.00 (optional)

Kills from previous years are recordable if mandatory data can be provided and verified. A trophy registration book will be available in the future.

Send recording fee and application to: NATIONAL BUFFALO ASSOCIATION, P.O. Box 706, Custer, SD 57730. Additional applications available.

TROPHY BUFFALO (BISON) REGISTRY

TABLE 1 MEASURING INSTRUCTIONS FOR TBR SCORE

A steel tape must be used and the horns should have dried 2 months or more. Measure to the nearest one-eighth of an inch. Enter fractional figures in eighths.

1.) Tip to Tip is the measurement between tips of horns.

2.) Greatest Spread is measured at a right angle to axis of horns.

3.) Length of Horn is measured from lowest point on under side over outer curve to a point in line with tip.

C4) Circumference of base is measured at a right angle to axis of horn. Don't follow the irregular edge of horn; the line of measurement must be entirely on horn material, not the jagged edge. Repeat for other horn.

C5-6-7) Divide #3 of longest horn by four. Starting at base, mark both horns at these quarters (same point on short horn) and measure circumferences at these marks, with measurements taken at right angles to horn axis.

Add it all up and you have your TBR score. To compare or convert to Boone & Crockett scoring system; subtract 'Tip to Tip' and 'Greatest Spread' measurement and symmetry differences from TBR score.

		SCORE
1.	Tip to Tip	= _____
2.	Greatest Spread	= _____
3.	Length of Horn right left	= _____
C4.	Circumference base right base left	= _____
C5.	Circumference 1st '¼' right 1st '¼' left	= _____
C6.	Circumference 2nd '¼' right 2nd '¼' left	= _____
C7.	Circumference 3rd '¼' right 3rd '¼' left	= _____
TOTAL.		= _____

BUFFALO

APPLICATION SUBMITTED BY: SCORE MEASURED & WITNESSED BY:

Name _____   Name _____

City _____ State _____ Zip _____   City _____ State _____ Zip _____

Signed _____ Date _____   Signed _____ Date _____

* NBA/ABA Sanctioned

----PLEASE TURN OVER AND COMPLETE INFORMATION ON BACK OF FORM----

19
Proven
Promotional Ideas

There's nothing better for promoting anything than buffalo.

Because of their novelty and the romantic associations in the public mind, buffalo grab attention, draw crowds and inspire enthusiasm. Buffalo have served well as the publicity focal point of town centennials—founders' days—of launching community projects—of charity drives—grand openings—you name it—it's happened already, and successfully! A live buffalo in a pen (high, strong, tight, vandal-proof: you don't want a bewildered buffalo charging down Main Street!) in the town square draws people to anything.

Just about any kind of promotion goes better with a buffalo. You can raffle off a real live one . . . or make him the Grand Prize. Anything that needs public attention benefits from a buffalo—live, cooked, or frozen. Buffalo BBQ's are surefire fund raisers. Offer as prizes buffalo robes, heads, skulls, or meat. Another approach is to set them up as premiums or bonuses.

And there's this curious thing about the buffalo mystique—you don't even need a real live buffalo to draw a crowd. Their dried droppings do just fine.

PROMOTING WITH A BUFFALO CHIP FLIP

"Let's start a World Buffalo Chip Toss Olympics," Jennings idly joked back in '72 at a meeting to plan a new magazine, *BUFFALO!*

Nothing much came of it. Pierre, SD, picked it up for Pierre Days, but it remained for Judi Hebbring, NBA's go-git'm Executive Director, to put some creative organization and full-steam promotion behind the idea. Pretty soon Buffalo Chip Flips were a necessary part of all manner of community celebrations. People write and call NBA for instruction and advice on staging their own hometown Chip Flip. For their edification and success, Judi composed the following instructions, commentary and rules:

Staging Your Own Buffalo Chip Flip

When spring is in the air, thoughts turn to your own hometown Buffalo Chip Flip. Here's a quick education on the fine points of this newest Olympic-class sport:

Buffalo Chip Flipping is an excellent media event to publicize a fund raiser for charity or other worthy causes. Publicity and promotion people term a "media event" anything unusual or spectacular enough to attract coverage by the news media. Always handle the subject lightly; it's all good, clean fun. The light approach attracts media. Billing your event as an "organic frisbee throw" or "Bushwhack County Buffalo Chip Flip Olympics" or "Interplanetary IFO (Identified Flying Objects) Launching Contest" or some such cutesy title helps to bring the reporters and the camerapersons. It's a good idea to work "buffalo" in there somehow.

Some years ago I organized such an event to run with Custer's annual summer celebration. (My attempt to get the longest buffalo chip flip entered in the *Guiness Book of World Records* succeeded only in getting them to drop their cow chip category. Darn!) I issued news releases, invited known record holders from other events, and POP!, calls started coming in from all over the country. The wire services picked it up and I taped numerous telephone interviews for radio stations coast to coast.

On Chip Flip Day, camera crews began arriving. PM Magazine from Denver covered us. NBC had a crew on hand taping for a New York TV station, as well as for possible use on Real People. Area TV news crews arrived, as did newspaper reporters.

Most of the contestants were locals, but several record holders came from a distance. A few tourists heard about it on the car radio and decided to come to Custer to try to become famous. One family heard about it while traveling through Illinois and even had appropriate T-shirts made to wear while competing. Many of these people did on-the-spot interviews with the media.

For days after, I did more telephone follow-up interviews. I was overwhelmed with the interest this thing generated, even though this was my first. So you can see, if handled creatively, the potential for publicity is unlimited.

Fund Raiser

Publicity and more publicity is all the more important if you're flipping chips to raise money for a new hospital or something. The larger the crowd, the bigger the take. Sources of income include entrance fee, sale of the chips, and glove rental. Of course if you're doing this only for fun or publicity, you don't have to charge anything.

Distance

My first few chip flips were distance events. We roped off a course about 50 feet by 200 feet, placed 25-foot markers, and added to the hilarity by measuring distances with two sticks and a lasso.

If you're allowing multiple throws, it saves a lot of measuring if you wait until the contestant completes his/her throws, then measure just the longest. It helps a lot to have a nimble-footed youngster run the results to the scorekeeper.

Hot Bodies Needed

A well organized distance Chip Flip needs at least six people. Two stretch the rope, one measures, one is a runner, one a recorder, and one to register and control contestants. Either the recorder or the registrar needs a loud voice or a mike to announce the results and to keep the throng apprised of the intricacies of the contest, throwers' individual techniques (underhand, overhand, roundhouse, discus style, shot-put, Australian, freethrow, etc.). Having a few willing extras around is a good idea, as runners wear down pretty quick and sometimes the field people get fed up with being splattered by chip chunks and chuck it.

Precision

If accuracy is your aim (PUN INTENDED!) rather than distance, this is the simplest way to go. You will not need a lot of space, help or makeready. Just draw a circle about 12 feet in diameter with chalk, paint, or use a garden hose. The hose is best, as it can be moved closer to the throwing line as people tire, or if you have several categories of contestants—such as Romper Room, Macho, and Sunset Set. Provide a bullseye in the shape of a tire, paper plate, etc. All the help you need is someone to keep track of who gets closest to the mark and someone to write it down.

For most, the excitement comes from getting up nerve enough to pick up that first chip. You need a way for them to get rid of it, so you have them throw it and it's nice to have an objective such as accuracy or distance.

Categories

Dividing your competition into classes draws more contestants: Adult men, adult women, teens, children and tots. The children can be further divided by size or sex. There should be a winner from each class and a grand champion overall winner. You could have categories for people on crutches, in wheelchairs, or on stretchers, even. Prizes are up to the discretion and imagination of the organizers.

Homemade trophies add to the fun, such as chips spray-painted gold, silver, bronze or mud and mounted on boards for plaques, hung on blue ribbons with medallions, etc. Chips have been varnished and fixed up as paper weights, desk pen holders, wall decors—only your imagination limits what you can do at zilch cost.

It's also fun to have a special award for the overall worst flip of the day. This usually goes to some macho guy who's trying to impress his girl and ends up making a fool of himself with a foul chip (out-of-bounds, we mean).

A Chip Flip gives a buffalo rancher a golden chance to garner a lot of free publicity by making a big deal of his providing the Official Competition Ammunition. How he takes care that they are 100% unadulterated, genuine, pure buffalo chips; how they are gathered, dried, stored, guarded from tampering (so no one can load a chip with rocks), etc.

Probably the most preposterous and hilarious aspect of any known Chip Flip came one rainy summer when it was impossible to find hard, sun-dried chips to flip and the secretaries of the local Chamber of Commerce office stayed up late the night before trying to dry them out and firm them up with hair blowers!

Help is at Hand

If you need additional guidance or copies of the Official Rules, we'll be happy to assist from NBA Headquarters.

RULES AND REGULATIONS
GOVERNING BUFFALO CHIP FLIPS
UNDER NBA SANCTION

1. Only pure, unadulterated, 100% organic buffalo chips will be used. Domestic cattle chips, beefalo chips, and hybrid chips are unacceptable. There will be no distinction made between bull chips and cow chips.

2. The buffalo producing the chips will not be fed cement kiln dust, heavy grains, buckshot, or any other product that could alter the weight of the projectile.

3. No adding or including heavy objects, such as rocks, will be allowed.

4. No chip under three (3) inches in diameter will be used in competition.

5. Each contestant is allowed to choose his own chip from a common stockpile, first-come, first-served. In the event of a crowd, chips in good condition may be reused.

6. Chips will not be eaten.

7a. Distance flipping: The object of distance flipping is to hurl, throw or otherwise propel, without the use of artificial devices, a buffalo chip, as far down the prairie as is humanly possible without injuring anyone

!!!!!!!!!!!!!!!!!!!!
Buffalo Chip
Flipping
!!!!!!!!!!!!!!!!!!!!

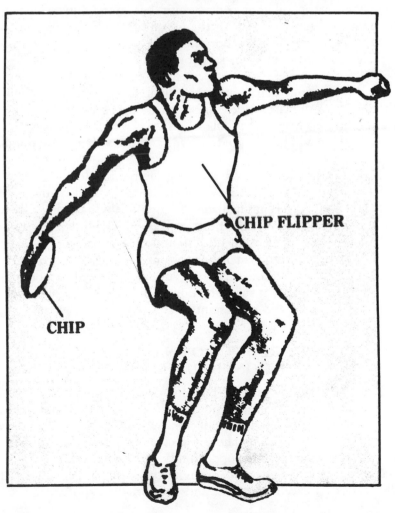

CHIP FLIPPER

CHIP

... including one's own self.

7b. Accuracy flipping: The object of accuracy flipping is to sail, skip or otherwise propel a chip, as near to or within a designated mark, not more than 40 feet nor less than 20 feet from the point of origin, as possible, without splattering the judges with debris.

8. There is no rule 8.

9. Number of flips allowed each contestant is up to the discretion of the judges. Generally, we recommend two (2) flips per contestant.

10. Flipping out of bounds, i.e.: into the crowd, at judges, at TV cameras, or at other contestants will be severely penalized.

11. Flipping out of bounds because of ineptitude will not be penalized. Determination of intent will be at the discretion of judges.

12. In the event the chip deteriorates in midair or on contact with the ground, the largest remaining piece that lands will be judged.

13. Chips that roll upon contact with the ground will be judged from the point of first contact with the ground (or whatever).

14. Contestants will be allowed to flip in any manner that is comfortable to them. Manner or style of flipping will in no way prejudice the judges.

15. All decisions by the judges are final. Disagreements may be met with a chip to the face. Rules may be made up or altered by the judges at any time, even in the middle of a flip.

16. No one will be allowed to parade around with a "chip" on his/her shoulder.

17. Wadding or in any way altering the natural form of the chip is strictly forbidden and all throws of this manner will be disqualified.

18. Licking fingers immediately following the handling of chips is not forbidden; however, the practice is discouraged.

19. Chip flips are recommended to occur following, rather than prior to, a meal, unless sufficient hand washing facilities are provided.

20. Gloves are discouraged. Persons using gloves will be labeled "sissy" and may be penalized at the discretion of the judges and subject to ridicule by the crowd and contestants.

DECLARE YOUR OWN COUNTY-WIDE NATIONAL BUFFALO WEEK!

The concept of a National Buffalo Week declared by the NBA got off to a good start with numerous buffalo producers around the nation tying in and promoting their product in a variety of ways. A special week to promote buffalo raising and products is going to work only if producers take advantage of the opportunity and promote regionally. Those who made a serious effort to promote the week of September 11–18, 1982, reported excellent results and good media exposure.

As the concept grows, we expect to receive national attention, but for now we'll dwell on successful, localized promotions that may help you plan for future National Buffalo Weeks.

Several buffalo farm "open house" events occurred on specified days during N.B.W. Door prizes were given and samples of the product were offered as well as tours of the facility. This sort of event drew attention in the area and introduced potential customers to the product. This worked especially well in non-traditional buffalo raising areas.

A number of producers gained media attention by simply filling in the blanks on the press release issued by the NBA office and providing their local media with the information. No special events were scheduled, but a number of newspapers ran stories highlighting the producer's buffalo operation and thereby brought attention and possible new customers with very little effort.

Our favorites among the information provided by those who promoted N.B.W. both occurred in Iowa. I suggest those of you in the other 49 states take note.

Ramona Laubscher of Denison, IA reports: "As soon as we received the letter and news release about National Buffalo Week, I went to work.

"First, I visited with each of three of our local restaurant owners and asked if they would be interested in helping us promote buffalo meat during the week. All gave me an affirmative answer immediately. They each made plans to put ads in the local paper for each establishment's weekly specials and entertainment. This they did so the public was made aware of the meat being served and when each restaurant would be serving it. They each told me how many pounds of buffalo they thought they could use.

"Secondly, we delivered the meat along with a large poster and dark brown 2-ply napkins to each of them on the Thursday before the 11th. Each poster was 22" by 28" on colored poster board with the name of the restaurant, pictures and dates on them. (I retrieved them to use again next year.) They were great advertisement and eye catching! The napkins had a large side view of a buffalo and NATIONAL BUFFALO WEEK — LOCALLY SPONSORED BY LAUBSCHER ACRES on them. We had the napkins printed locally and they were a good source of advertising too.

"Third, during that week, Jimmie and I went to each place to eat. They each served something different and reported the public was very well pleased and they sold out far sooner than expected. One manager said some people came a day early because they wanted to be sure they got to taste it, but had to return to dine the following evening. The restaurant managers told us they would like to do it again."

The Hartmanns of Everly, IA, tied their promotion in with the Clay County Fair as National Buffalo Week coincided with the fair dates. Each Clay County farm commodity is given a special day for promotion during the fair and this year, with N.B.W. coinciding with their events, a day was set aside to promote buffalo products in the Farm Bureau's booth. Jim manned the booth, which was decorated with a head mount and other "buffaloey" things and passed out samples of the product. The weather on buffalo day wasn't very good, but Jim reports a lot of promoting was accomplished even under the adverse conditions.

For added media exposure Hartmann presented KICD-FM radio country music DJ, Rick Friday, a buffalo chip for lunch on buffalo day at the fair. Friday, who was broadcasting from the fair grounds, is very diet conscious, and since his show runs over the noon hour, he generally says something like, "What's for lunch, bunch? I'm going to have an old dry donut," or something like that. The radio station staff helped set up the buffalo chip for lunch presentation without Friday's knowledge. The station manager thought it was such a good idea he insisted it be done live on the air. Hartmann reports that Friday, who is never lost for words, was caught a little short that day. Friday was a good sport about it all and afterwards he and the radio staff enjoyed some buffalo meat provided by Hartmann.

Hartmann tells us that KICD has been very cooperative with giving NBA air time, especially around our convention dates. Jim's advice to NBA members is if you're not getting news releases to your local media, you're missing the boat. Media people are looking for the out-of-the-ordinary and buffalo aren't found on every farm and ranch.

For those of you who are new to the association, National Buffalo Week is declared by the NBA, not by Congress or the President. We had a week declared by Congress once, but the declaration wasn't approved till the week was half over! Then we later learned that each special week must be re-declared annually, must carry a tremendous number of co-sponsors and must go through the entire committee approval process before becoming a fact. We, as an association, decided that it would not be in our best interests to be bothering our Congressmen and Senators with this situation every year. We need their ear on far weightier matters than that.

We simply declare our own week. With a national organization making the declaration, and regional promotions occurring, we expect our week to gain acceptance as we develop our promotions.

One effort that would give the week better clout would be *state* declarations of N.B.W. This you can do by asking your governor or state legislature to sign a proclamation or resolution declaring "Buffalo Promotion Week" or whatever. This will give you media attention in your state and will probably get you on TV and in the newspaper as you look over the governor's shoulder while he signs the proclamation.

MORE FREE PUBLICITY

Make use of all the publicity you can get. Buffalo are still scarce enough to be news, especially in areas distant from traditional buffalo territory and near metropolitan areas. If you work the media right, you may never have to pay for advertising. Let your local media people know you are doing something unique in your area. Gathering news isn't as easy as you might think and they are always looking for something a little different with local flavor.

Invite area TV and radio stations and newspapers out to your place for a visit and a sample of your buffalo meat. Media food editors are especially good targets for this approach. Your returns of free publicity are going to greatly outweigh the free samples you give out. Let them know whenever you are doing something special, like sorting or branding or butchering—you might get a story out of that. Have an open house and let your media know in advance. Put up a buffalo display in your local museum. Approach local restaurants with the idea of adding buffalo to their menu as an occasional "special." They will probably receive such favorable input they'll want to carry buffalo as a regular menu offering.

CATCH THOSE SPECIAL EVENTS

Big ticket, fund raising dinners—like political parties, hospitals, and other charitable enterprises put on—need to offer something super special on those high priced plates, something more exotic than roast beef and mashed potatoes. Buffalo is the answer, of course!

The Friends of the Museum, Milwaukee, served themselves buffalo and charged themselves $125 a plate for it. They paid Dick Kuhn, Lazy K Farms, $2 a pound for two carcasses on the hook.

You might as well cash in on the Christmas rush—everybody else does. Dick reports one raiser makes extra monies by packaging nine-pound gift boxes of various cuts for the Christmas mail order trade. Sezze he, "It's not the least expensive way to market, but the best way to get the highest price per pound." The bottom line is what counts.

Leading
Buffalo
Raisers

V
V
V
V
V
V
V
V
V
V
V
V

20
Talks
With Outstanding
Private Ranchers

FROM HAND-MILKER TO BUFFALO BARON

Whether the Durham Ranches comprise the world's biggest buffalo ranch sorta depends on where they are in their annual marketing cycle. Durham spreads over 88 sections of semi-arid bunchgrass and sagebrush along US 59, 30 miles south of Gillette, WY. Just before roundup, they'll have close to 3,000 head. After they sell of the surplus bulls and old cows, the census will be closer to 2,000.

Durham Ranches are owned by Durham Meat Company, San Jose, CA, a wholesale purveyor of fine meats to the institutional trade and the military. Over the past 20 years, Durham Meats has become famous for consistent high-quality buffalo meats. They have also built a loyal following of fine hotels, restaurants, clubs, and exacting gourmets.

The owner and president is Armando Flocchini, an immigrant boy who never went to high school but proved America is still the land of opportunity for those with grit and gumption. Armando is short, stocky, and distinguished, with a thin line, precisely sculptured white mustache. He became NBA's vice-president upon its founding in 1966, again in 1981, and served as its president in 1983. His trim and charming wife, Lena, often accompanies her husband on his frequent visits to the ranch, where they spend about four months a year. She is always seen with him at NBA functions.

They built themselves a sumptuous log house atop the highest butte to give a clear view of their far-flung spread. "Besides," Armando grins, "the wind blows the snow off!"

Armando's son, Bud, is vice-president and general manager. He has his own log house on the ranch, some distance from the Big House. Second and third generations are active with the ranch work; Bud spends about two months a year on the ranch. A grandson lives and works on

Mr. and Mrs. Armando Flocchini.

High atop a windy hill, the Bar D big house commands
a sweeping view of the entire spread.

the ranch year-round. Great grandchildren are still more noise than help, but they're coming along. Five families live on the ranch year-round.

"We thought of building the houses all together, like an old world patriarchal villa," Armando smiles. "And then we thought there would be too much noise." Son Richard is secretary-treasurer and plant manager.

Jack Errington, familiar to any NBAers who have attended meetings, is assistant general manager with more than a quarter century's service to Durham.

Besides managing the ranch, a 12-million-pound cold storage plant in Lincoln, CA, plus the San Jose packing plant, Armando oversees a 6,000-acre farming operation in conjunction with the ranch. While ranch and farm gross a million a year, this is just a spit in the bucket of Durham's $40 million annual gross.

Armando Sr. was born in France of Italian parents who brought the boy to Humboldt County in California, in 1913. He grew up hand milking, as you can feel by his grip today.

"I got tired of milking cows and decided to find something else to do." He went straight from grade school to business college and finished the commercial/bookkeeping course in eight months, then became an instructor. He was so young that no one else would hire him, despite his having trained himself by correspondence courses to be a railroad steam engineer and auto mechanic.

In 1926, he became an expert calf skinner. Armando explains that in those days of uncertain refrigeration, veal calf carcasses were delivered to butcher shops hide-on to preserve the bloom. Butchers didn't like to skin them, so Armando went from shop to shop, skinning the veals and buying the hides. He could skin 150 a day, traveling from store to store. His 60-hour week earned him $12.

In 1932, still a skinner, he borrowed $5,000 unsecured from his boss. With this he bought into Durham Meat Company, established in 1884 and then consisting of four employees. Over the next 20 years he bought the rest of the company and expanded into ranches running 5,000 cattle. In 1963, to collect a debt, he acquired 500 buffalo. Soon he bought the ranch land, too, which is today the Durham Ranches, running buffalo under the -d (say Bar D) brand.

"It started as kind of a hobby and then I made it a subsidiary of Durham Meat Company," says Armando. They've replaced about 65% of the fencing, still building five to ten miles a year (fences total some 200 miles!). They built two new log homes—one 3,000 square feet, the other 4,600—and the two big flat storage metal buildings, each 40×120 and 60,000 bushel capacity.

The semi-arid area gets eight to eleven inches of moisture most years. The land was infested with sagebrush and greasewood, but three years of aerial spraying brought that under control.

HOW WE MANAGE 88 SECTIONS AND
3,000 HEAD OF BUFFALO

by Jack Errington
Assistant General Manager
Durham Meat Company

Durham handles the buffalo herd twice a year. We round them up in September to brand the calves and dehorn the females and sort off the dry or open cows. They then go to slaughter, and we sort out the killer bulls.

Don't Use Horses! We do all our gathering in vehicles. Buffalo and horses don't like each other, so it's best not to work buffalo a-horseback in close quarters. We used horses at first, but in our big pastures we found the buffalo had more stamina than our horses, which gave the buffalo the advantage. So we switched to using pickup trucks and other vehicles because we could handle the buffalo better. We are fortunate in that our ranch is mostly rolling plains and easily accessible by vehicles. Don't prod buffalo. If you start prodding them, you can't get them to move!

In December or January, we put cows and calves into separate pastures for weaning. We found you don't run buffalo in adjoining pastures due to their family style of life. A buffalo cow will pace the fence, just the same as cattle, grunting for her calf, and the calf acts just like beef calves, grunting and trying to get back to its mother. But it doesn't last as long.

Pasture Rotation

We separate animals by ages. Yearlings come out of weaning pasture in April, into a large pasture until October or November. After the ground is frozen, they graze winter wheat until March, when we put them into the two-year-olds' pasture.

The twos have a large pasture divided into north and south sections. We rotate by placing the incoming animals into the north section the first year, which enables the south section to rest a full growing season. The next year the incoming twos go into the south section while the north section rests. This rotation doesn't deplete our grass. The threes are handled the same as twos.

The big herd has three pastures; winter pasture is the same every year. The other two are rotated the same as above to give each pasture a season's rest.

Weaning

We wean calves in December and January. We hold them in corrals for two or three days with hay fed on the ground. Then we put them into a larger corral and start them on small pellets (the size used for sheep) fed in five-ton barge self-feeders, with hay on the ground. Then in 10–15

Cattle guards save a lot of gate opening. Bar D guards are designed to discourage buffalo migration.

days, we run them into a 2,000-acre pasture of which 400 is stubble field. The weaned calves continue to receive the pellets but we reduce the hay amount since they pick up part of their roughage requirement from the standing stubble. We hold them in this pasture until April, when they go into the yearling pasture.

We figure 36 acres of pasture per animal unit without supplemental feeding. We actually stock heavier than that because we supplement feed during winter. (In summer, the buffalo don't eat all the tall grass; they save some for winter!)

We feed a standard cattle liquid supplement in all fields year-round, fed in vats with licking wheels. Certain parts of the country are deficient in certain elements and liquid supplement manufacturers can formulate a special liquid to overcome these deficiencies, but that is not necessary here. Liquid is fed year-round to all animals free choice. The feed is designed to limit the average intake to 1–1½ pounds per day, depending on the season. The 500-pound-capacity metal vats have two licking wheels each. The animals rotate the wheels through the liquid and lick the feed off the wheel. Calves on their mothers learn to use the lickers before they are weaned. This makes the weaning a lot easier. The liquid feeders go near the watering places in each pasture and that is when the buffalo use the lickers—when they come up to water. One advantage of the lickers is that the feed is available to the buffalo during storms when we can't get out to feed.

We feed small pellets from 5-T sheet metal self-feeders mounted on pipe skid for moving to new locations when the animals tromp out the dirt in front of the feeder.

The pellet contents vary depending upon grain prices. Last year's pellets were made up of 10% molasses, 35% milo and 55% barley with 1,000 units of Vitamin A per pound. We buy them from the local co-op, furnishing our own barley. (We raise 2,500 acres of wheat and barley each year plus 800 acres of hay.) The co-op delivers the pellets in auger trucks that unload directly into the feeders. We feed these pellets during December and January when the ground is frozen so the truck has no problem getting to the feeders. The weaners will consume approximately five pounds of pellets per day during the three- to four-month weaning time.

The buffalo regulate their own ration, consuming more supplement when grass is short. When there's a lot of grazing, they don't use the lickers. In winter, we feed range cubes and hay.

We feed cubes every day and hay every other day. The animals get a daily ration depending on age, size and weather. When we put out cubes, we feed them a double portion. They will clean this up the same day so there's no waste. The hay is fed every other day; in addition, we put several stacks in each pasture, which allows the animals to consume more hay if needed. It also makes hay available to the animals when it is too stormy to get out with our equipment. Buffalo don't tear up a stack and trample the hay as cattle do. They just eat around it until there's nothing left.

Our hay is mostly dryland alfalfa, which gives us only one cutting. We have a few small fields next to Hay Creek which are subirrigated and we normally get two cuttings, rarely three. We stack about 1,200 ton a year and some years we have to buy more.

Native grass is western short grass and buffalo grass. We reseed farming fields, when we switch from wheat to barley to hay, with two pounds crested wheat grass per acre, two pounds of intermediate wheat grass, two pounds of brome grass and four pounds of alfalfa. For pasture improvement, we disk in six pounds of crested wheat, two pounds of clover and two pounds of alfalfa per acre. (Our custom feedlot operation is covered in the chapter on "Feeding.")

Antelope Play A Lot Here

When you make your ranch good for buffalo, you make it good for wildlife. As many as 2,000 antelope roam the ranch. They can't jump our fences, so have to wait for us to open gates to change their range. When they become too thick, they damage the grain fields by eating the grain and just by walking through the standing grain.

Pumps are timed to pump 15 minutes an hour. This keeps the water fresh in summer and unfrozen in winter.

Water

Buffalo can travel farther from water to grass and back than cattle can. But we try to provide plenty of water tanks to avoid overgrazing near water while grass goes to waste farther out. Most of our wells are equipped with electric pumps set on timers. We set the timer to run the pump 15 minutes each hour. This keeps the tanks from freezing, and the overflow forms a nice pond, allowing more animals to drink at a time.

Breeding

When we roundup to wean the calves, we pregnancy test the cows by the arm test: The veterinarian inserts his arm into the rectum and feels the calf. The arm test is valid after three months' gestation. We do it at four or five months and it is about 95% accurate. We don't give open cows a second chance. When we were building the herd, however, we did. We don't keep breeding cows over eight to nine years. They breed in summer pasture in July or August. Heifers breed at age two and drop their calves at three.

Bull Management

Nature provides a bull calf for every heifer, but this is a luxury no working rancher can afford. We experimented, using two-year-old bulls on two-year-old females at a 1:1 ratio and got 95% pregnancy test by the vet. Prior to this experiment, we used to put our two-year-old heifers in with our breed herd just before breeding time and usually got only 80% pregnancy rate upon test.

We sort out the top 10% of our two-year-old bulls after they have been used on the two-year-old heifers. These are added to our breeding bull herd in September after we have sorted off the old bulls, any cripples, or any bull that hasn't fully developed into a type we are looking for. We plan to have most of our breeding done by three- or four-year-olds, when they are most active.

Our main breeding herd will consist of, as a rule, 40% threes, 30% fours, 20% fives, and 10% sixes. We sort off approximately 10 animals out of each age bracket in September for slaughter, except for the 20 head of fives and sixes that go into separate pasture as hunter bulls. By doing this, we find that the number of old bulls that are usually run off by the young bulls is greatly reduced because we retire our bulls at six. Another benefit is we do not have the trouble now we had with the old bulls that used to tear up our fences and go wherever they wanted. Once they learn to break fences, you can't keep them anywhere.

In our herd, you'll find 90 herd bulls; 15% of them will be scattered out by themselves and not doing any work. Rather than pasture a lot of idle bulls, we move them out, else they get smart and learn how to tear up your fences. That's why we never put bulls in adjoining pastures with only a fence between. They will fight each other through the wire and tear up your fence.

We sort out the bulls in September, keeping them separate until we

Durham Meat Company fences prevent a lot of buffalo absenteeism.

Sturdy welded pipe gates ride on dolly wheels to prevent sagging.

go into winter pasture. That way, we don't have to fight those bulls to round up the herd at weaning time. Bulls run with the big herd except from September to winter. They go back into the herd after the calves are weaned.

Inbreeding

We make no attempt to avoid inbreeding. At first we brought in outside bulls, but we get better results using our own. We've got some big bull power out there. All but the top 10% we move to the slaughter pens.

Dehorning

We dehorn all heifer calves because we operate just the same as a commercial livestock operation. The livestock industry, except the purebred operators, has dehorned both male and female cattle for years. Horned cows rip each other up in the corral and in trucks and spoil a lot of meat. Cows become so irate when herded into the corrals, they often gore and kill their own calves! You don't lose as many calves when the cows are dehorned.

We dehorn at four months with a scoop dehorner, then sear with a hot iron to kill the horn bud cells.

237

Bar D alleys prohibit animals from jumping out.

Connecting rod between gates opens one when you open the other for sorting.

Branding

We use four electric branding irons: One is the -d brand and the others have three numerals each: 1 2 3, 4 5 6, and 7 8 0. We invert the 6 for 9. During the 1980's, the year digit goes over the brand; then in the 1990's we'll put it under the brand. Since we don't keep an animal beyond five years, we know from the brand exactly how old it is.

Labor

The ranch is home to five families year-round. The farming and haying require more hands than would a straight ranching operation. You don't need as big a work crew if you're just running buffalo. We use six hands year-round, add two or three during haying and another three or four for roundup.

Health

BRUCELLOSIS: The herd was already quarantined by both the state and federal governments when we took over. We knew we had a problem and prepared to clean up the brucellosis right away. We built corrals so we could institute a testing program. Our first test showed about 35% reactors. Stock was scarce and brucellosis in buffalo does not cause abortion, so we made arrangements with the federal government to run two herds—one clean, one reactor—on opposite sides of the highway. After five years, we slaughtered the infected herd.

NO MORE VACCINATION: We started vaccinating the calves and found we were getting a high incidence of reactors from vaccine titre. Some people believe the test reacts not to the disease but to the antibodies produced by disease or vaccination, while others vaccinate their calves without problem. On our last test, the only reactors had been vaccinated as calves. Whether vaccination was causing the reactions or not, we went to Washington and got the right to NOT vaccinate. We got our certification in 1968 when we achieved a negative test and release of quarantine. Since then, all our tests of killed and live animals have not shown a single instance of brucellosis. We've got a clean herd.

PARASITES: Parasite problems are minimal and usually require no treatment. We have some dust bags around the watering areas, as buffalo won't use oilers. We have little trouble with worms. They all have a few lungworms but fecal tests don't show any need for treatment. Some years we do have to worm a few yearlings and two-year-olds.

DISEASE: No disease problems have appeared, except occasional pinkeye, which we doctor a little.

Marketing

We have three classes of killers:
The primest, after custom-feeding, go by truck directly to custom

Mounted on a tractor front-end loader with hydraulically adjustable wings, this Bar D machine crowds buffalo down the alley.

slaughtering in or near San Jose, and are delivered in quarters to our plant. The only animals we slaughter directly off the ranch are the dry cows that are sorted off during our September roundup. These animals are normally very fat and acceptable for fabrication into steaks, roasts, etc.

We sort our cows again when we wean our calves. At this time we pregnancy test each cow and remove the open ones for slaughter. We also sort any cripple or undesirable cows at this time. These cows are put on feed for 60–90 days before slaughter. All of our two-year-old bulls are placed in a feedlot prior to slaughter. All cull bulls are also fed before slaughter. Trophy bulls are reserved for hunters. (This is covered in the chapter on "Hunting.")

BREEDING STOCK: Most of our sales of breeding stock in the U.S. are by private treaty. One of us deals directly with the customer. We require a 20% deposit to hold animals. Then we finish all tests and papers required by the state of destination. We require payment in full by cashier's check or money order before the animals leave the ranch.

We held a public auction in 1981, offering approximately 350 head, but sold only about half due to a poor turnout of buyers. The annual auction is being reconsidered.

CANADA: Our northern neighbor is developing into a good market for breeding stock. In the first three years, we sold a major portion of our breeding stock to Canada by private treaty, and also engaged a representative working on commission. We use Canadian truckers, as they seem to have less trouble getting through customs.

In any event, use only truckers who've had experience hauling buffalo. Don't use double-deckers or possum-bellies for mature bulls. Those on the upper deck poke holes in the roof with their horns. Those on the lower deck destroy the upper deck, causing those on top to fall in on them. Mature bulls should be hauled in a straight truck. Don't crowd 'em!

Buffalo won't load against the sun. You can't even drive them against the sun. They also won't unload at night. The best way to unload buffalo is just open the tailgate and walk away. They'll unload themselves.

MEAT MARKETING: Durham Meat Company's business is selling primal cuts of beef to institutions and the military. At one time we were breaking 300 carcasses a day, but boxed beef competition has cut that to 150.

We process buffalo meat into commercial cuts like retail portions: ribs, sirloins, fillets, etc., like beef. The hump is too gristly for steaks. We think the "hump roasts" you read about in the old days were really rib eye. The hump is best used in the ground portion.

We sell some carcasses direct from our place of slaughter. All other carcasses are brought to our San Jose plant for fabrication. The fabricated buffalo meat is then placed in our freezer for further processing as orders are received throughout the year. Most of our buffalo meat goes to gourmet customers built up over the years. Some have bought from us every week for years. Clubs that want to put on special feasts are another ongoing source of sales. Several regular customers buy buffalo meat on their doctors' recommendations because they can't eat beef.

PRICING: We do not base our price by what the current beef prices are, because beef prices change every week. Our prices are mainly based on demand for each particular item. We have used the same prices for two years or more. Buffalo, being what Armando calls "a romance type of meat," carries a premium. By making our production consistently high in quality, we have been able to maintain a price averaging 150% of the same beef cut. Even our trimmings get a 15 to 25% premium, although we have had to sell some at the same price as beef trims.

PAST PRESIDENT ESTABLISHED FIRST FARM HERD IN NEW YORK

Charles Tucker, Jr., NBA president 1981–82, established the first buffalo herd in New York. Home base is one of the few bicentennial farms in the state. Homestead Farms was purchased from a Madam Brett land grant in 1743; the original indenture parchment hangs in the parlor. Charlie's grandchildren are the 12th generation to live on the farm at Stormville, which encompasses 1,500 acres operated by Charlie and his two sons. Homestead Farms is also a beef ranch with a few sheep on the side.

pilots for quite a while, wound up flying 4-engine transports full of gasoline from India across the Himalayas into China, surviving 69 trips. Today he's a member of the Hump Pilots' Association.

In the fall of 1945 he came home with his wife, Julia, and found pilots were a glut on the market. So he stayed home and milked 24 cows with a crew of five. Milk went down and feed went up, so he decided to get big and efficient. In 1951, he build a 310-foot pole barn with 64 stanchions. But efficiency couldn't make up for low prices, so in '63 he switched from dairy to beef. Steers didn't pay, so he tried stock cows in '65.

That summer he had the thought that when you have something no one else has, you have a valuable product. Half in jest he started looking for buffalo, finally finding some on the Sutton ranch at Agar, SD.

"Believe me, the education has been vast, extensive, and expensive!" NBA's *Buffalo History and Husbandry* hadn't been written yet so he had to learn the hard way! NBA was still a year away (Charlie became a charter member).

Homestead Farm at Stormville, NY, was built by Issac Storm in 1750.

One of the major problems Charlie encountered with his fledging herd is common to the east: Chokecherry leaves. Says Charlie, "As long as they are green, no great problem. But let a strong wind or some buffalo that wants to scratch come along and you have a broken branch with green leaves on it. The wilted leaves are full of prussic acid and an enzyme that actually eats the lining of the stomach and intestines. It kills in about

242

Charlie Tucker, past NBA president.

Charlie didn't start out to be a buffalo farmer. He trained to be a commercial pilot and graduated from Parks Air College with a BS in aeronautics in 1940 when we were gearing up for war. He instructed

24 hours. The only cure was to wait until winter (when there were no leaves to wilt) and cut down all the cherry trees we could find. Even now if we have a bad storm, we ride around the pasture to see if any new ones have broken off." Charlie lost 20 head to cherry leaves and 35 to parasites before he got these things under control.

"I had tried worming them in the feed with all sorts of remedies, and everything you can do, outside of physically catching them. Nothing worked. The strong ones got the medicine and the weak ones, which needed it, got nothing. Then I put in a Thorsen chute and started worming them individually. That stopped my losses from parasites.

"As a suppressant, we mix phenothiazine 1:12 with salt and keep it in front of sheep and buffalo all during the parasite season. Levamisole hydrochloride is an oblet (a big pill) which we actually put down their throats twice a year for stomach worms, intestinal worms, and lung worms—it catches everything.

"Due to the chance of the parasite becoming immune to phenothiazine, we take it away during the winter. However, I suppose in the south you would have to use it all the time. It's just a suppressant to keep parasites in check until you worm with tramisol. Tramisol by dart gun is fine if you keep track of which ones you have done.

"I would not change this information one little bit. It took me 17 years to come by this. When I started, buffalo were represented as being so tough they could stand anything. Much to my sorrow and expense, I have found this to be true only when you have 15 or more acres per head to roam on as out West!"

NBA PAST PRESIDENT PRACTICES WHAT HE PREACHES

NBA past President Lloyd Wonderlich and his son, Richard, run 50–60 head on their 500-acre corn, hog, cattle and soybean farm near Mt. Pleasant in Iowa. They feed all the grain they raise to 500 fat cattle and 500 hogs; just a little goes to their buffalo slaughter bulls.

Lloyd, a native Iowan, has farmed this place about 30 years. Dick graduated from Iowa State in animal science. "Then we made him go it on his own for a couple of years to see if he really wanted to farm, and he couldn't wait to get back." They formed a family corporation, Wonderlich Farms, Inc., and "we both work for wages."

Lloyd and Dick bought their first buffalo, four heifers and a bull, in 1970. "Richard and I like the challenge. When you handle buffalo, it is a challenge." They find that by starting with calves, good cattle fence will hold them. "Once you get this established as their home, they'll stay. But you turn a bunch of old cows out and they'll keep right on a goin' through every fence in the country!" Corrals, of course, have to be stronger and higher than cattle handling facilities.

NBA past president, Lloyd Wonderlich and son, Richard.

Basically, the Wonderlichs just let the buffalo roam on 50 acres of blue grass and orchard grass pasture, seeded in the spring. "After about 10 years, it will be back to blue grass, which is a very good grass for buffalo. If you have good moisture, two acres is sufficient for one cow."

They run one bull for 10 cows the year-round. Doing what comes naturally, the bulls automatically breed the cows to calve in April and May, the best time for buffalo. They do not separate the bulls, except those bound for slaughter. They drive out in the pickup every day and throw off some corn to them "just to keep them gentled down a little. Now, whenever the truck drives through the pasture gate, the buffalo come a runnin'. Makes 'em easier to lead into the corral that way, too."

Buffalo refuse to enter any shelter, preferring to head into any blizzard that may happen by, baby calves and all. Most winters, buffalo merrily paw through snow to get grass, but some winters seven-foot drifts are too much even for bison. Then Lloyd and Dick bulldoze a trail into the buffalo pasture and feed hay—70 half-ton round bales, all told. Says Lloyd, "I don't feed the cow herd unless we get a foot of wet, heavy snow which they can't forage through. The calves are not fed grain until they are weaned.

"We're going to adopt a practice Roy Houck has been preaching for years: Give 'em minerals free choice and they'll balance their own instead of feeding them a mineral mix."

The Wonderlich's herd gets handled but once a year in November when they are corralled, the calves weaned and the surplus bulls cut out for finishing and slaughter. They don't get much finish—just a little corn on the ground.

Marketing is practically automatic. Every year they could sell more meat than they can produce, just by word-of-mouth. Of course, other breeders buy three or four bulls a year.

They shoot the fat bulls in the lot and haul the carcasses to a nearby state inspected locker plant as needed to fill orders. The locker plant cuts them up, packages, and quick-freezes the meat. The Wonderlichs keep the meat in a walk-in freezer on Dick's place a quarter mile east of Lloyd's. "Most of our meat goes to people who want a gourmet main dish for family get togethers. One town in Illinois buys 3,500 patties every year for a community celebration. (We find 'chopped buffalo' sells a lot better than 'burger.') Another community buys two whole buffalo every year for some kind of whing-ding."

Then, every year over the Labor Day weekend at the Old Threshers steam reunion in nearby Mt. Pleasant, the two families set up a circus tent, barbeque four whole buffalo and sell over a thousand sandwiches every day of the five-day event. Says Vivian, Mrs. Lloyd, "There's GOT to be a better way of chopping roast buffalo than with scissors!" Mrs. Dick, Esther, pitches right in. Diners like it so well they buy packages of frozen meat to take home.

"Dick took a picture of a mother and her new calf which was printed in FARM WIFE along with a little story about our buffalo. We could have sold a hundred buffalo from that one little article."

There's a lively market for buffalo by-products, too. "We sold $400 worth of horns, green hides, and stuff at a muzzleloader's rendezvous. They pay $10 each for the horns to be made into powder horns. They tan the scrotums for bullet pouches. They carve scrimshaw on the toe-nails for jewelry. I keep a bucket of toenails to fill orders." Lloyd boils the meat off the heads in a big old lard kettle in the barnyard and gets $30 per skull. Boiled horns slip right off, adding another $20 on the head's value. Green hides for mounting fetch $75. He has a standing order for all the winter killed hides he can produce at $60, and summer hides are worth $30 green. People turn the hooves into bookends, lamp stands and gun racks. They make fly twitches out of the tails. "There's no end to the buffalo market." Wonderlich, who served NBA as President from 1979 to 1981, sees NBA's big job right now as getting regulation to label cheap Asian water buffalo as such, so people won't think they're getting good American bison buffalo meat in stores and restaurants, when they're actually getting the vastly inferior water buffalo, which feels like card-board in your mouth and is twice as tasty.

NBA has never promoted buffalo as a get-rich-quick venture. I asked Lloyd if buffalo made any money for him and Dick in their first decade.

He stroked his weathered chin and looked at the puffball clouds clabbering up for another spring rain. "Well, now, we're limited to the size of our herd by the amount of pasture we've got. We're thinking of seeding down more pasture and expanding our herd. Buffalo helped pay for the cattle these last few years. Buffalo have been awfully good to us."

STANDING BUTTE RANCH

Triple U Enterprises

One of the most remarkable buffalo operations is the Standing Butte Ranch owned and operated by the Roy Houck family on the west bank of the Missouri. It is about 30 miles upriver from Ft. Pierre in central SD. Its size alone is enough to make it remarkable: an irregularly shaped spread of 50,000 acres measuring 10×18 miles across the extremes, with 2,000 to 3,000 buffalo. It is one of the world's largest privately owned herds. Cattle, horses, camping, buffalo, a tourist ranch and a slaughterhouse, which processes buffalo exclusively, all operated under the UUU brand. But 'twas not always thus.

The Houcks scrabbled through the dust bowl and Depression by strategems ranging from milking cows to trailing their beeves overland to grass in Nebraska and putting up hay for sale when they could find some. Then the Oahe Reservoir took their ranch on the east bank of the Missouri up by Gettysburg. With the reparation money, savings and credit, they bought the present spread in 1959, overgrazed and run-down.

Roy Houck, patriarch of Triple U Enterprises.

247

"We had this new ranch and there was a lot of room and we thought buffalo would do good," Roy recalls. "Buffalo always had an appeal for us and we had a longing for them, wondering how they'd handle." They bought 23 head in Montana and picked up odd lots here and there. Then in 1963 they bought 500 head from Custer State Park in southwestern SD. They were in the buffalo business!

Years of careful conservation ranching, understocking, and a full time dam building project made three or four blades of grass flourish where one used to struggle. Networks of ponds and slow running artesian wells prompt Roy to boast, "No animal, wild or tame, has to walk more than half a mile for a drink on this ranch." It is not unusual to see buffalo, antelope, and mule deer grazing together.

The ranch now has a carrying capacity of about 3,000 head, more or less, depending on rainfall. As their buffalo herd grew, they reduced cattle numbers to match the land's productivity. It was not until 1973 that the annual increase permitted regular slaughter. Then they built a Grade-A packing plant right on the ranch, fitted with the best sanitation equipment and huge freezers capable of quick-freezing a roomful of the huge, steaming carcasses.

For slaughter they put their surplus 2½-year-old bulls into drylot for a 90-day finishing feed, just like cattle. The result is juicy, well-marbled meat more suited to the American taste than grassfat, although some connoisseurs prefer the latter.

Way out in the rolling grass hills, 30 miles from town and 200 miles from the next packing plant, the abattoir is run entirely by family members. Sons, daughters, their spouses and a grandson are all working ranch crew. The plant supplies supermarkets, fine restaurants, and special orders by air express, frozen. The plant also custom slaughters for other buffalo producers. Other packing plants are reluctant to process buffalo because of the beasts' rambunctious nature. And federal regulations make them close down and clean up before resuming cattle slaughter because buffalo are wild animals. (Other federal regulations consider buffalo domestic animals. It's all very confusing. NBA is working to get the buffalo's legal status clarified and established.)

Roy's length, breadth, and depth of buffalo experience qualify him to be called The Dean of the buffalo business. He was founding president of the National Buffalo Association, serving from 1966 to 1976. As president, he was instrumental in launching the NBA magazine, *BUFFALO!*

BUFFALO IN VIRGINIA

by Wray S. Dawson

The 225 acres we now call home is named Thistlewood Farm. It is approximately 35 miles from Washington, DC, and 25 miles from our store in Annandale. It was a run-down place, but we fixed it up and moved our Angus herd from our former farm. We moved into our rebuilt house in 1975.

I had always been intrigued by buffalo and had been trying to find out about raising them. In 1973 we bought nine cows and a bull from Bill Neff of Harrisonburg, VA. After I got into raising buffalo, I sold all my cattle, much preferring to raise buffalo.

For four years we sorta did things by trial and error, as I am sure many buffalo people do. However, it was a great day when we found out about the National Buffalo Association. We joined and went to our first meeting in Florida in 1977. We were delighted with the organization and all the wonderful people. We have been very active with the group ever since and I was indeed honored when I was elected to the Board of Directors and given the job of membership chairperson, then marketing co-chairman and recently vice-president of NBA.

It soon became obvious that I would be needing a corral and some way to handle the animals for separating and shipping. I knew my cattle equipment would not work. I had seen an ad for the Thorson headgate in *BUFFALO!*, so I sent for information and found that Charlie Tucker had one. I went to see Charlie's layout. He was very helpful and gave me a lot of good pointers. I then called Jerry Houck, who was also very helpful. After I had seen the Houck corrals, I was ready to come home and finish mine.

The first catch pen is 150×150 with 4-gauge, 5-foot galvanized bull wire set one foot off the ground. I used locust post 12 inches in diameter and 9 feet tall. In one corner I put a gate to be operated with ropes from a distance. This gate opens into an alleyway 150 feet long which leads into the crowding alley. The entire setup is made from nine-foot telephone poles and rough cut oak. Everything is super strong and I did most of the work myself with the help of my tractor and front end loader. The corral was finished in November, 1978, and we attempted to have our first roundup.

This did not prove to be too successful as we were using tractors with front end loaders and the animals were being cut up too bad. After putting about 10 animals through the headgate and getting two in the pen which were going to be shipped away, we had a large cow get into the alleyway and some way got herself turned upside down. What a time we had getting her out of there without breaking a leg or killing her! We had to take the sides of the alleyway apart. Was she ever glad to get out!

We called a halt to our first roundup and I did a little redesigning of the corral. I also made a piece for my front end loader to cover all the sharp edges so now they don't get hurt if they bang into it. We have learned, as many of the other NBA members, that the less you handle the buffalo, the better they get along. I give my animals a little crushed corn about once a week just so when I call them I can get them to follow the tractor because they know I have a front end loader full of goodies for them.

In the spring of 1978 I purchased 18 buffalo from the herd that Arthur Godfrey had on his place near Leesburg, VA. However, I only got 13 of them as they killed five in the process of getting them delivered. They had no facilities for loading the animals. They used a tranquilizer gun and then roped them to get them into the trailer. As buffalo people know, this is a definite NO, NO! First of all, the tranquilizer was not good and second, the roping probably choked them to death. Another lesson learned the hard way.

We have only begun to sell a few of the animals the last couple of years. We have sold a few live calves and the past three years we have furnished a local hunt club with two buffalo each year for a "buffalo roast."

We have had a great deal of publicity. It seems that the novelty of our place is the truly western atmosphere of our farm which lies only 35 miles from Washington, DC, in the Blue Grass Hunt Country where some of the greatest race horses in the country are raised. I love the buffalo and will continue to enlarge my herd as much as I can. I feel there is really a great future for the meat in this area as well as the many by-products we derive from the animal. I've gotten in on the ground floor of what may be the wave of the future. I'm committed to the buffalo business even though it was eight years before I showed a profit.

These animals consume nothing usable by man. Right now, cattle feeders feed five pounds of grain that people could eat to produce one pound of beef. With the bison, you use only grass. With all the ballyhoo beef breeders make about special breeds, they can never match us on this one point alone. Domestic cattle have to be fed grain to produce marbling for quality meat. As world population increases, we'll be faced with serious food shortages and we're not going to be able to put much grain into animals. That's why I have such great hope in buffalo for the future. And buffalo meat qualifies for a health food.

Like most buffalo meat producers, I've had to create my own market. I sell wholesale only, a side being the minimum quantity sold. It is priced about 40% over beef. I usually sell to individuals for their home freezers, declining to serve restaurants because I can't guarantee a steady supply yet. The demand is so great we seldom eat it ourselves, even though we prefer it to beef. We reserve our buffalo for special occasions.

If you're going to raise buffalo for money, you've got to be your own entrepreneur. I've never had to advertise. My meat is sold three years ahead. All I have to do is pick up the phone and call. People are ready and waiting. Most producers sell all the meat they can raise to friends.

We butcher in December and January when the pelt is prime. The skins are made into robes which sell for $450. Nothing is wasted. Local

Dawson with his specially equipped front-end loader.

Indians, following tradition, take the skull, hooves, and even entrails for making ceremonial objects. Especially prized is the skull. A well-utilized buffalo brings $2,000.

An ordinary farm fence will hold them unless they don't have what they need in your pasture, then they'll break out to get it. A buffalo isn't going to stand there and die like a cow.

My herd will follow me like I'm the Pied Piper. But if there's a stranger with me, I can't handle them. You can't drive buffalo. You can only lead them.

A cow with a calf is something else. You'd better not get between them. People have driven a car between a cow and her calf and the cow tore the car to pieces. She was defending her calf. If an incident happens, there's always a good reason. Most of the time it's the foolishness of the people.

BUFFALO IN VERMONT

by Barbara Pettinga

The large dark shapes bunched behind a low red barn aroused casual curiosity. Were they cattle or were they horses? The curiosity became more than casual when a closer look showed them to be neither. Buffalo in Vermont? Indeed! Fifteen of them; cows, calves and two heavy caped bulls—their shaggy heads appearing awesome and exceedingly impressive to anyone expecting to see their domesticated cousin, the cow. Fifteen acres of hilltop farmland in Charlotte, Vermont, overlooking Lake Champlain and once the eminent domain of dairy cattle and sheep are now home range to these Plains bison.

"Cattle are neuterous animals," said owner Tony Perry whose first four six-month-old calves, three heifers, and one bull calf, came from Valentine, NE, via Doc Miller, a New York livestock dealer. "Buffalo are infinitely more interesting than cows and horses," he said.

Although he concedes that many of his reasons for having bison are emotional, Perry is experimenting with their practicality as meat animals for New England farmers. "When I moved to Vermont, I noticed farming was on the wane, particularly the small family farm," he said. "For farms to exist, especially high mountain farms which I call horse farms because they're difficult to work with a tractor, you've got to have animals. I tried cattle and couldn't even get my hay money out of them." Deciding he might be more successful with a subsistence animal accustomed to foraging in the wild, Perry turned his first four buffalo calves out with cattle on a farm he owns in Danby, Vermont, in 1974. He later moved them to his Charlotte home for convenience of management and so that he can enjoy watching them.

During the first half of the 20th century roughly 80% of Vermont was open land; the remaining 20% forested. Now the tables are turned

252

with fields and pastures accounting for a mere fifth of the land area. Many people lament a passing of the pastoral scene which was typical of the state's recent past and they seek economically feasible ways of reversing the small farm abandonment trend which is resulting in woody plant succession and a subsequent lack of scenic variety.

Perry bought six more buffalo including a mature bull and cow. "I waited four years for the first calf which was born in 1979," he said. Five were born in 1980 and Perry was there to observe the birth each time.

"I'm not interested in developing a large herd. I'm more interested in seeing if raising them in Vermont is feasible. Most of what I know about buffalo I've learned on my own," said Perry who feels this approach is realistic since most Vermont farmers could afford neither the time nor expense of attending seminars, workshops and so forth. "The down side is the fencing, although I'm convinced that if you put them in a large area with plenty of food, you can confine them with a four-strand barbed wire fence," he said.

Perry's fifteen acres are divided into several pastures fenced with five strands of barbed wire. "DANGER KEEP OUT" signs are conspicuously placed. One large holding area, easily viewed from a bay window in the old wood frame and brick farm house he is renovating, is enclosed by heavy gauge, five-foot-high mesh topped with two strands of barbed wire and fastened to five-inch cedar posts spaced ten feet on center. Sometimes the buffalo get out, either by jumping over or crashing through the barbed wire fence, but "they don't stray very far."

"I grain them for control," said Perry explaining that they'll come when he bangs the grain bucket. He said he has no trouble loading them into a specially designed and reinforced trailer because he keeps it parked at the pasture gate and baits it with grain for several days ahead of time. "I've butchered two animals so far," he said. One was a three-year-old bull that walked onto the trailer the day he was scheduled to be taken to a butcher for slaughter, ate the grain and lay down, according to John White of Dorsel, Vermont, who has been Perry's hired man for sixteen years.

"I had to kill a yearling heifer who got a foot caught in the fence. I didn't like the meat as well as the three-year-old bull; it didn't have as much flavor," Perry said.

Although in winter bison can satisfy much of their need for water by eating snow if water isn't readily available, Perry's herd is supplied by "an old bathtub with a water heater and water running in it all the time."

Bales of hay, kept dry in the barn, are fed to the bison every few days in the winter. Last summer the grasses in their own pastures provided adequate forage for them during the growing season.

Perry expressed concern that ultimately a diet of Vermont vegetation may not provide all the nutritional needs for these animals habituated to prairie grasses.

The only parasite problem Perry has is sore spots on the withers of certain animals in summer. In much of their historical range, wallowing in alkali soil is supposed to help, he said. "Their natural protection is rolling." When his buffalo quickly learned to avoid sprays directed at them and various hanging devices rigged to treat them with repellants, he mixed insecticide into a couple of truck loads of sand brought in and dumped in their pasture. "That seemed to do the trick," he said. "I worm them every six months in an unprofessional way, by pounding up medication and putting it in their grain."

"In the east," Perry said he was told by a New York buffalo raiser, "you've got to be careful of cherry trees." Browsing standing bushes or newly down branches is not harmful, but wilted cherry branches will kill them." Vermont state veterinarian, Dr. David Walker, confirmed that cyanide poisoning which results from ingesting wilted cherry bark, leaves, and seeds is a problem with any ungulates, or hoofed mammals. Several species of wild cherries, exceedingly common along Vermont roadsides, are among the shrubs cut down by highway maintenance crews, and partial wilting of these bushes is accompanied by chemical processes that yield hydrocyanic, or prussic acid. Although "cattle love 'em," according to Dr. Walker, cyanide poisoning is not very common among dairy people most of whom are pretty well aware of the potential risk. This type of poisoning occurs in summertime only, with death attributed to anoxia, or oxygen deficiency. Diagnosis is facilitated by the abnormally bright red (oxygen starved) blood of affected animals.

"Another problem I've got is a fruit orchard," said Perry whose apple and pear trees are enclosed with the buffalo, by wire mesh fences. "They check those trees out every day in autumn." He is afraid the animals will gorge themselves after a windfall and that some may die. Fruit juices which ferment in the stomachs of ruminants (animals such as cattle, bison, sheep, and deer that have four-chambered stomachs) cause a terrific intoxication according to state veterinarian Walker who said that cattle will eat all the fallen apples they can get. "It just about dries them off. They're all done milking for that lactation," he said.

Perry said he had been in the habit of going into the enclosure with the buffalo until "I had a once in a lifetime experience that changed my mind." It happened late in the summer of '79 when he had gone into the pasture looking for a wet spot because he wanted to dig a hole. For a moment he had his back to a bull and two cows, one of which he was tending a considerable distance away.

When Perry heard a rumbling sound he spun around to see his dominant bull only yards away and closing fast. The mere seconds involved seemed long as Perry's common sense told him to run. Instead he reacted aggressively by jumping up and down, yelling and waving his arms. At what seemed the last possible instant the bull veered off, ran by, circled and turned to face him again. Perry repeated the jumping, yelling, and waving his arms. Then, fearing a second charge, he began to walk "calmly" toward the fence. When he was close enough he "ran like hell" for it. "Then I was very human. I picked up clods of dirt and threw them at him. It was the worst situation I've ever been in," said Perry, whose history of close calls includes a head-on automobile collision and almost getting caught in an avalanche.

Literary accounts of bison have often depicted them as being awkward, stupid, and barely able to see out of their dull little eyes. Close hand observers, like Tony Perry and John White couldn't disagree more. John White said, "From the size of them, you'd think they'd be clumsy, but they look like they've got springs in their feet; they're so graceful." Perry finds moonlight nights especially good for buffalo watching. They do tag games, chasing each other and sometimes hitting on all fours like a mule deer. "They'll chase a dog just like a dog will chase a cat," he said. As for their eyesight, he believes it's comparable to an elk's or a deer's, animals generally thought to be alert and observant. Dull? Not a bit! "Their eyes sparkle," he said.

It's hard to imagine future scenes of Vermont valleys and hillsides darkened by buffalo herds. Historical evidence of their presence in New England prior to European settlement of the east is lacking, although, in a limited sense, their range probably extended to include parts of Pennsylvania and New York. Skepticism of the authenticity of Perry's herd is such that many local people, hearing of them, assume they must be beefalo (or cattle), a cattle/bison cross. Once they are convinced, however, that these are "the real thing," delight in having them is almost universal.

THE FINE POINTS OF ZERO MANAGEMENT

Right where the county road shoots off of WY 50, 12 miles south of Gillette, a signboard lists the ranchers along the road. Clear at the bottom is TOOTS MARQUISS, 30 miles.

You drive the gravel south, over rolling sagebrush country, bouncing through dry gulches and past wind carved buttes to delight the heart of any cowboy movie fan. Watching hawks and eagles make their majestic circles in the unpolluted blue, you may be lucky enough to see one make its bullet-like stoop to snatch up a rabbit or mouse. You'll likely see deer and antelope at play, expecting every new vista to bring buffalo into view, for this is buffalo country. You won't see any though, until you get to The Little Buffalo Ranch, Lazy S Hanging 6 Brand.

Meander a mile down a single track and you'll see a beautiful house of cut stone, knotty pine, and logs just across the Belle Fourche River.

Toots Marquiss, the Buffalo Baroness.

Ford the river (don't worry—it probably isn't deep enough to drown your engine) and there you are. Across the next gulch you see a cluster of ranch buildings, including son Gary's house, and hangars sheltering three airplanes. "Airplanes save a lot of time in ranch work," Toots explains.

Your first sight of Toots Marquiss—tiny, petite, with a clear, high, girlish voice—make you think "China doll." Yet, she's the Buffalo Baroness. You don't have to know her very long, noting the sun and wind etched face, firm grip, straight back and jutting jaw, to realize here is a ranch woman of pioneer stock who has weathered storms worse than wind and rain. "China doll," you decide, "of 24-carat stainless steel!"

This is one of the oldest private herds in the nation, roaming wild and free, and managed only as nature manages it. It is a hobby for Toots, her sons, Gary and Trigg, and her seven grandchildren. Her husband's father, Ted, bought two cows and a bull in 1922 out of the Scotty Philip herd, just for fun.

On a neighboring ranch lived a pretty little girl named Opal Elaine Wagensen, but her folks and everybody called her Toots. "If anyone says 'Opal' to me now, I never think to answer," she giggles. "I'm Toots. Always have been." The Marquisses "just always had buffalo, even when I was a little girl."

She married Quentin Marquiss and they began a family. When the children were in school, the family lived in Gillette and commuted to the ranch weekends to see how the hired man was getting along.

"Buffalo are like rabbits," observes Toots. "If you're not careful, pretty soon you've got too many." By the '50's, the herd numbered close to 500, more than their range could support without damage. They sold 25 head, to Denver. "The trucker was glad to get our business. He had this brand new truck and he thought the publicity would be real good for business, and he asked us to ship all future loads with him.

"Those buffalo butchered themselves before they ever got to Denver. What a mess! And his truck wasn't new anymore. He decided he didn't want any more of our business and we decided never again! Marketing live buffalo isn't all that shiny. We've never touched one since. They don't interfere in our business and we don't interfere in theirs."

When her husband died in 1964, Gary was 19 and away at aviation school, Trigg was 15 and Glo was little. "I don't know how to market buffalo. There's no established market like there is for livestock. Every time you want to sell one, you have to go out and make your own market. My husband understood that. He enjoyed that. But buffalo are difficult to handle. You don't just load them into a truck. You do all sorts of good stuff before they're ready to go." So they cut the herd to 75 head, "a nice hobby herd," she smiles. "We want it to remain a hobby. We don't really want it to be a business, but if you don't cull it, it turns into a business."

"How do you handle your buffalo herd?" we asked.

"We don't handle them at all. We have a 12,000-acre pasture with a six-foot fence around it. Buffalo are meant to be in their natural state. I don't think they were meant to be worked or corralled. It's a shame there are rules and regulations and brucellosis and all the monkey business. When we butcher, we just try to take out the ones that aren't quite as pretty. There's not much improvement you can make. We try to keep the herd at 75 adults, sorta."

The Gillette *News-Record* quotes her: "We just keep them as they were 'way back when.' They're a nice animal to have around, since they take absolutely no care. We don't do a solitary thing. You never lose a buffalo in a storm. It would be the business to be in altogether, because they're disease-free. At least our herd is." She reports their death loss is "nothing" aside from the few that roam their home until they die of old age.

"When we want some meat, we drive out into the pasture and shoot one. We field dress it and haul it to a locker plant in Casper for processing."

When they get a trophy bull who's outlived his dominance, driven off by a youngster, they advertise him to hunters, and haul them out in a pickup. The prince of Iran, Abrodezza Pahlavi, came to the ranch in 1960 to get a trophy head for his palace wall. A Harvard grad, he wanted something that best symbolized America. "He made a speech you wouldn't believe. It was better than anybody in America could have put out, I think, like 'the buffalo is America and it is the history of America.' He really knew American history better than the rest of us."

The herd is totally inbred and linebred. No new blood has been admitted since the original. "I've not seen any change in the herd in the last 30 years. I think we have as pretty a little herd as I've ever seen. They've not deteriorated at all through inbreeding."

The Marquiss family runs cattle, sheep, and buffalo together in the same 12,000-acre patch, a little more than half a township. Springs and reservoirs assure plenty of water, and the Belle Fourche River runs all year. "We don't feed the buffalo at all, in any way. They live and die right there. We put out salt for the cattle and sheep and I guess the buffalo lick that, but we never feed them. They'd get to be an awful nuisance when we tried to feed sheep or cattle in the same pasture because the buffalo do have the upper hand—ah—horn."

The ranch spreads through the rain shadow of the Big Horn Mountains that range the western horizon, cutting rainfall to as little as eight inches a year. Wheatgrass and buffalograss dominate, amongst the sagebrush, yucca and cactus. "It's a big advantage to run sheep, cattle and buffalo all in the same pasture. You get more total grazing because the sheep eat the low-growing vegetation and the buffalo eat the high, and they both eat stuff the cattle won't. The buffalo do beautifully just on what nature provides, burrowing through the snow or cropping the tops off the tall grass."

NBA recognized Toots by naming her to the Buffalo Hall of Fame in 1981, an honor reserved to those who have been of greatest service to the industry. Says Toots, "I've supported NBA right from the start." She's a neighbor to Armando Flocchini on the -d just two ranches away (Armando was a founder of NBA, its first vice-president, and became president in 1983). "I admire NBA's causes and dedication. It's a beneficial organization. NBA does the country as a whole a lot of good, not just buffalo ranchers. That's why I belong."

HOW TO SUCCEED IN THE TOURIST BUSINESS
(without even trying!)

(Born in 1901, Martin F. Collins is the oldest active NBA member that we know about.)

"When I was six or seven, my Daddy took me to see the buffalo. We went by train from our homestead in Lyman County to Yankton and Ft. Pierre and then by buggy to Scotty Philips' ranch. I remember looking out from the house in the morning and seeing the big old bull on the hill and something went on in my mind and I said someday I'm going to have some buffalo," recalls Martin.

Martin Collins at 80+ years
is still raising buffalo.

The family had moved by covered wagon just after the turn of the century from Wynot, NE, to claim a homestead in Lyman County, north of White River, SD—a property the family still owns. Martin married in 1925 and took his bride to a hotel room in Wynot where they lived quite a spell while he started his first buffalo herd. "She said we didn't own a chair or a dish but we could still buy buffalo."

259

Ride'm Cowboy!

One of his bulls, "Buff," earned respect around the rodeo circuit in 1925, '26 and '27 as a bucking buffalo. The easiest way to earn $10 in those days was to ride ol' Buff. Martin did it six times in one season, a sort of a record! Not everybody made such easy money. From cowboying to rodeoing he took to clowning until "I got busted up pretty good so I quit."

Martin got his schooling in Oacoma and Chamberlain, SD, taught school and worked in the Homestake gold mine while the family bought the Nissen homestead in the center of the Black Hills in 1926. You may have noticed Collins Buffalo Home on US 385 between Trout Haven and Pactola. The drought and hard times beginning in 1929 forced the liquidation of his Nebraska herd.

You Can't Keep The Tourists Out!

When, in 1940, times got a little better, Martin bought some buffalo from Custer State Park, animals tracing back to Scotty Philip. Collins Buffalo Home is on the main tourist route through the Black Hills, and of course passersby couldn't resist. "We didn't make any play for tourists, but we couldn't keep'm out!" He built a strong steel mesh fence where he keeps a few of his more peaceable animals so the flatlanders can feed them handfuls of hay through the mesh. "We have to give them hay or people say 'don't you feed them anything?'" Also on display are Yaks, Longhorns, Scottish Highlanders, and some crossbreds.

"The bus drivers make a big thing of the buffalo so their passengers insist on stopping. The kids are the main thing—it's hard to get them to leave. One bus driver told me he hates to come in here . . . takes him so long to get the people away."

How To Whip The Worms

Martin holds his herd to about 50 adults, rotated on four pastures. "We use range cubes as a teaser. For fattening, we give a gallon of medium ground corn. Shelled corn and ear corn go right through them. We use NBA's first book, *Buffalo History and Husbandry*, like a dictionary."

With buffalo spending so many months in the little lot by the highway, worms are a problem. "But I get Dr. Wendell Peden out from Rapid City and he gives Tramasol and they snap right out of it."

Marketing

For butchering, he puts two animals in his trailer with a gate between to keep them from skinning each other up and hauls them to a locker plant. There the meat is packaged into family size portions and frozen. He sells the frozen packages out of his home freezer.

"Buffalo are our sole support. We make no gate charge but we did put up a donation box. About half the people put something in. A bus-load will leave anywhere from $5 to $40. However, our greatest compensation is the pleasure we get—meeting and conversing with people from all over the world. We receive letters and cards from Scotland, Paris, Switzerland, all over. Foreigners are hot for anything buffalo. That's their hobby, playing cowboy and dressing Western. Our last sale of skulls and robes went to France and Switzerland."

The whole family is involved in the Collins buffalo enterprise. Smiling Mother Collins answers the phone and greets guests. Daughter, Mary, keeps a remuda of saddle ponies and posts a sign by the gate, "Mary's Rides."

"I am very willing to stay at our home and promote buffalo by talking to several thousand visitors yearly from every state and many foreign countries. I extend an invitation to anyone traveling through the Black Hills to stop and visit our small, exceptional herd.

"Just having the wonderful animals here to live and work with brings our better than 80 years of life to a wonderful finish. It has been a long, hard and sometimes discouraging struggle from bucking Buff and his wild associates to what we have today."

You can tell by Martin's grin he feels like a rich man.

21

Top
Buffalo Herdsman:
Fred Matthews

(*Editor's Note:* When we started writing NBA's first book, *Buffalo History and Husbandry*, back in the early 1970's people said, "Be sure to talk to Fred Matthews, Custer State Park buffalo herdsman. Hardly anybody knows more about buffalo than Fred." Our intentions were noble—our efforts mighty—our performance zilch. We can plead only the greatness of distance and the thinness of available time. We hope this chapter makes up for our former lack. Custer State Park is unique in being self-supporting. The buffalo herd contributes a big part of that support, an endorsement to Fred Matthews' herdsmanship.)

LONG, TALL TEXAN

Tall and spare as a Texas' rancher ought to be, Fred Matthews was born about 100 miles south of Dallas on June 22, 1917 into a ranching family. The Matthews also ran a meat market, doing their own butchering and marketing the meat from their ranch. Prime cuts went for sale and the family ate the stuff that most people didn't . . . like brains, kidneys, tongues, and livers. Fred to this day gags at the thought of eating liver.

He worked with his father during the Depression, helping put brother and sister through college—a luxury Fred never had. When Fred was 19, his father died and Fred worked on ranches until he was drafted in 1942.

Stationed two years near Rapid City, SD, he married a local girl. From 1944 to 1946 he pushed across France and Germany into Berlin. Back in South Dakota, Mrs. Matthews planted roots and started their family. For nearly nine years, Fred worked on a ranch near Hermosa on the east edge of the southern Black Hills near Custer State Park.

Fred rodeoed a bit in the '40's and '50's. He got pretty good at bull riding and nearly attained some championships. It took only one bad accident, however, to convince him that bull riding was too risky. After

Fred Matthews.

he healed, he stayed off the bulls, switched to bulldogging and then to calf roping.

Although he liked ranch work, the long hours and low pay left him dissatisfied. Through proximity to Custer State Park, he became familiar with the Park operation. In September, 1954, he hired on as a skinner, since butchering was one of his sharpest skills. The regular hours, improved working conditions and better pay agreed with Fred. In 1982, after 29 years of service, he was still working at Custer State Park.

BUTCHERING

When he came to the Park, there was no corral system and no means of cropping the buffalo herd except through slaughter. The Park sold all excess buffalo as meat. Quantity meat orders came from all parts of the country. They'd butcher 20 head a day, four days a week, after the weather turned cold, until orders were filled. There were no limits at that time to the herd numbers and no set number to be culled. They just killed as many as they needed to fill orders. Since the herd numbered 1,500 to 2,000 head, the buffalo kept well ahead of the annual cropping.

Aggressive culling in the 1960's helped produce Custer State Park's splendid herd. *Bruns*

CULLING

Fred simply picked out the animals to be slaughtered out in the open where he could get a good look at them. He always chose for slaughter the lesser animals with imperfections, leaving the best to propagate. This method of culling through those years is one of the prime reasons, in Fred's opinion, that they have such a high quality herd today. He wishes they still had this option, as it's nearly impossible to cull animals accurately by looking at them through the rails of a chute. "Field slaughter is the best way to manage a herd for improvement," he insists..

WHAT FRED LOOKS FOR IN A GOOD BUFFALO

NBA: "What is a good buffalo?"

FRED: "Oh, I look at the color and conformation, the deepness of the body. I think maybe there are two kinds of buffalo. We used to try to cull out those light bodied, light colored ones when we were field killing, and any that were small boned or had anything wrong with them. But you can't see all that in the chute. This is where the guy with 10 or 15 head has the edge. It's easier for him to see exactly what he's got and get rid of what he doesn't want.

"If you have animals with imperfections or poor dispositions, you get rid of them. You don't want them passing on those traits in your herd. People have been managing their herds this way for a long time even though there has been no registration. I think registered herds would speed up improvement. I also think we're all getting closer to having registered quality herds. With few exceptions, everybody is managing their herds to keep the good and get rid of the poor. That's what I started doing down here in the 1950's.

"When I started, some buffalo here had awful big humps on them, but their rear ends were small. The humps didn't look normal because they were so out of proportion. And you could see it in the offspring. I didn't think they looked good so I just started culling them out.

"We also had what we called 'rabbit leg.' That's the sort of thing that comes up in linebreeding. They'd have crooked hind legs set way forward. I'd watch that real close and cull them out. We haven't had that for a long time. Culling in those days helped this herd a whole lot."

CORRALS

Live auctions brought equal revenue to the Park with a lot less expense and trouble, so CSP gave up slaughtering. But auctions require handling facilities.

For his first year, Fred was strictly a skinner. Then Park Superintendent Les Price told Fred to design and build a corral system. "I don't know anything about livestock," he told Fred. "You do. You build 'em."

Fred had nothing to go on for his design except experience and

Part of Custer State Park's corral system. *Bruns*

Squeeze chute and headgate. *Bruns*

Why buffalo corrals and working systems have to be hog-tight, bull-strong and horse-high! *Bruns*

common sense. They picked a site in the south Park and Fred started out with four rodeo type chutes. He added to it, made a lot of mistakes, threw out some of it, added some more, and kept making changes until he got something that worked. Fred's corral design has stayed basically unchanged through the years, with some additions to speed loading-out at the annual sale. Ranchers come to copy Fred's design because it works.

They use a WW squeeze chute for the adult animals and a small Powder River chute for the calves. The WW is basically a standard factory model; a top to discourage the animals from jumping out is the only real modification for handling buffalo. Since CSP doesn't handle bulls over two years old, the chutes work fine. The old bulls are too big for the chutes, but "We don't put old bulls through the sale because people are afraid of them and won't buy them. They're nothing but trouble to us so we don't even try to round them up. When they're 10 years old they go to a hunter. That's the easiest way to get rid of them. We keep about 10 bulls out there in each age group from two years on up. This way, we've got 50 or 60 bulls working."

The corrals enabled the Park not only to sell excess animals at live auction, but also to start a selective breeding program. As soon as they had handling facilities, Fred started aggressive culling. The corrals also gave Fred the means to clean out the brucellosis. Although there was no law then requiring it, the Park decided to clean their herd voluntarily for the comfort of their cattle raising neighbors and to produce more salable buffalo. This decision in the early '60's turned out to be a good one with many benefits.

REGISTRATION?

Fred wanted to cut out about 50 of their best cows and a few choice bulls and isolate them in an area of their own. He thought to keep this herd closed, never selling off any heifers, and starting a registered herd from scratch. He would have gradually eliminated all the grade animals in the Park, restocking with the registered herd. This plan was never started, but the methods used produced a clean, improved herd anyway.

"I advocate registered stock, seeing that registration motivates owners to better management and faster herd improvement. The record keeping that goes with registry assures cleaner bloodlines and the time and expense involved spur people to strive to improve and swiftly eliminate any animal that could put a black mark on their reputation.

"But I don't see how Custer State Park could ever have a registered herd. For the small breeder, or even the large breeder keeping a small registered herd separate from his main herd, it would work to clean up the breed. I'd like to see CSP have a registered herd but we couldn't keep pedigrees on them. People should have them certified, though, so someone with a low percentage hybrid couldn't sell it as a pure buffalo even though it looks like one. There should be something to make sure it's pure buffalo they're dealing in."

LINEBREEDING

"A lot of closed herds are linebred herds. Custer State Park's is a line-bred herd. All of Custer's buffalo trace back to the original 24 animals introduced 60 years ago. Not a drop of new blood has come in. And we don't have any two headed calves or five legged cows out there. It all goes back to culling."

PARASITES

NBA: "Describe your parasite control program."

FRED: "CSP has no parasite control program. We had a guy checking that once. He found some indication of parasites in the droppings, but not enough to do anything about. We feed Flyblock and have for a long time. This helps keep parasites under control.

"Putting wormer in the feed works for some buffalo, but others have to be treated some other way. Some of them won't get to the feed because of the pecking order and others just don't like it and won't eat it. I wish they'd put all this stuff in range cake. Our buffalo all like the cake."

FEEDING

"The calves take right away to the creep feeders. This isn't to say, though, that you can't take an adult animal from CSP home and teach it to eat from a bunk. By now, all those adult animals out there have had some experience with creep feeders as calves.

"We feed the cake on the ground. I take a scoop shovel and give it a big fling to spread it out. If I put it in a pile, a few will hog it all, but if I spread it out real good, they all get some. When they hear that shovel rattle, they come a runnin' to the pickup. This helps us move them around. They'll follow that pickup with cake in it just about anywhere you want them to go. If you're trying to drive them, though, forget it!"

WEANING

NBA: "Tell us how you wean."

FRED: "We used to wean the calves out there. A cow carrying a calf is going to be in better shape if she doesn't have last year's calf sucking her. The calf she's carrying is going to get a head start and be a better calf.

"We'd separate them in January and keep them corralled until the grass was green, four or five months. They did real good, too. We'd feed them over the winter and then they'd go out on the good grass and develop real good. Even the late calves, the runts, would do good.

"Also, if she's sucking a yearling, that yearling might just kick that new calf off. Leaving the weaning up to the cow is hard on the cow

269

because a buffalo calf is hard to wean. I've seen them old cows trying to get away from those calves, just galloping right down the trail, and that calf right behind her, his head stuck between her legs, sucking while they both gallop along."

LATE CALVES

NBA: "You mentioned runts. What causes calves to come late?"

FRED: "Late calves, like lower calf crops, just seem to run in cycles. I don't think it has anything to do with what we did or didn't do. It just happens sometimes."

BRANDING & MARKING

NBA: "Tell us how you identify the animals."

FRED: "Before we had corrals, the buffalo weren't branded or marked in any way, so we had no way of knowing what age balance we had. Now, with handling facilities, we brand with a hot iron and tattoo and ear tag. We don't lose more than half a dozen tags a year. When we first decided we wanted identification, we tried horn branding.

"Up to that time, there was no management, really. We didn't know what we had because they weren't identified. We did some culling through slaughter and that's all. Now we know what we have because they are all identified and we keep an age balance as part of our management.

"Lately, we've been branding the little late calves with a smaller iron. This causes less stress to the smaller calf, but mostly we do this because sometimes those brands grow with the hide. If you start out with a big brand on a small calf, you may end up with a huge brand on the adult. This doesn't happen all the time, but it happens often enough that we try to protect against it."

WORKING SEASON

CSP used to round up and auction in late January or February. Now they do it in early November. Both seasons have their advantages. "I think the January or February sale gave us better conditioned animals to sell than the November sales. Snow on the ground eliminates the dust and pollen irritants. Working them with no dust eliminates some dust-caused problems. And, later in the season, the animals are in better shape all around. The calves have a better start. On the other hand, the heavy winter hair made it harder to read the brands."

Fred noted, "We had more pinkeye in '82 than usual. Our vet thought this was because of the tall grass that year and more pollen—more direct irritation. Or maybe it was the Mt. St. Helen's ash-fall."

HOT PRODS

NBA: "What do you think of hot prods with buffalo?"

FRED: "There's a time and a place and you've got to use your head. Then they're all right. We use them. As for when, how and how much, that's at your discretion. When a buffalo gets on the fight, nothing you do is going to help and a hot prod will make it worse. But when that critter gets started in there toward the chute, a little buzz will make him go. If you don't he'll back out and then you'll have trouble getting him going next time. If you overuse the prod, then that animal is going to be crazy when he hits the chute."

FENCES

NBA: "CSP must have a heap of fences."

FRED: "Over 40 miles of fence. You never catch up fixing fence. Last year we put in over 1,200 posts. Sometimes we just patch things up the best we can, but you can do that just so long and it catches up with you and you have to fix it right. Better to keep ahead of these problems and do it right in the first place.

"Buffalo are smarter than a lot of people think. They learn fast. Out in the Park, we don't have a problem with them wanting to get out. They know when they've got a good thing. They've got 50,000 acres of good grass, so they don't bother the fences."

UNPREDICTABILITY

NBA: "Would you say buffalo are pretty unpredictable?"

FRED: "I'll say they're unpredictable! I know how far to trust a buffalo and after that I don't know what they're going to do or how they're going to do it. I've heard people talk about 'tame' buffalo and doing things like setting kids on the back of a buffalo, but I wouldn't do it. Even a 'tame' buffalo isn't tame and I wouldn't trust one of my grandchildren on one.

"People go into the middle of a herd to take pictures. I don't doubt you can do that, because mostly they don't give a damn, but why take a chance? Because they are unpredictable and maybe today they do care. Buffalo don't like to be touched and I wouldn't trust one even if it's a trained 'tame' one. Not with a child, anyway.

"The most stupid people I've run into are photographers. They think they can get them to pose and it just doesn't work. During the roundup we have photographers all over the place. They don't care how bad they screw things up as long as they get what they want on film. I have a reputation of being terrible mean with photographers, but if they get in my way when I'm trying to work, I just put them out of the way. And you know, buffalo sometimes like to play and sometimes they get in a playful mood with photographers and this scares the hell out of 'em.

"We had this Italian film crew out there shooting a motion picture. They were all dressed up like cave men and there was this one guy running around in the buffalo herd and acting like he was scared for a movie

Buffalo are unpredictable. Don't test them.

scene and spooking the hell out of the herd. Well, about the third time they tried to shoot that scene, one cow had had enough. She started blowing snot all over the place. They said the leading man was muscle-bound and couldn't run very well, but he ran real good that day. He was the guy who was supposed to act scared. He didn't have to act scared this time. No one had to tell him that old cow was mad at him, either. He knew that instinctively.

"I've found I can do a lot with a whip without ever touching them. When I'm close-working them, I find the pop of the whip right in front of their nose will turn them."

BUFFALO NATURE

NBA: "I notice you seem to have no fear of those buffalo when you're in that sale ring with them. Is this because you can read them, know what they're going to do?"

FRED: "Yes, I pretty much know what they're going to do. It comes from years of working with them and it isn't something you can teach. I've found they'll almost always veer off even when they're coming at you with their tails in the air. But when they're standing there a-shiverin' and a-shakin', you better beware. If they've got a way to get away from you, they'll take it nine times out of 10. But remember, also they might not, so you can't take any chances. I can usually tell, though.

"One time when I was standing in the ring, an old cow got in there

Fred has little fear, but lots of respect when working buffalo.

273

somehow when my back was turned. They hollered 'Look out!' and I didn't have time to go anywhere and she was coming right at me. If I'd moved one way or the other, I would have probably moved right into her path. I stayed put and she veered off; I felt her horn just brush me.

"A lot depends on whether you're moving one or a bunch, and on the wind and whatever. If the wind is wrong, they don't want to go. If they're going to go, they'll move into the wind. If they want to go that direction, they'll go, even if it means running a horseman down.

"A buffalo is a playful sonavagun. He'll play with anything—a motorcycle or vehicle or anything."

REACTION TO LIGHTNING

"I don't see a lot of difference in working them in different weather conditions. But you can tell if there's a big storm coming: They'll split up into little bunches. This may be nature's way of protecting them. If they stayed in a big bunch, lightning could get them all. But the buffalo scatter out and they don't all get hit. Only one time have I found buffalo—two or three head—that looked like they'd been hit by lightning. The elk get hit a lot more than the buffalo."

VISION & SMELL

NBA: "How good is a buffalo's vision?"

FRED: "Medium. At a long distance, they either can't see or they don't give a damn.

"But they sure have a sharp sense of smell. A few years ago, three bulls got out and it was like they sniffed the air and decided where they were going. They acted like they were working their way toward Bear Country, a wild animal park 30 miles away near Rapid City. We were able to turn them OK, but we couldn't catch them and they kept doubling back. When they were nearly there, I took my 30-06 and that was the end of the chase. They seem to have smelled those other buffalo or something. We had a yearling bull go thataway one time."

TRANQUILIZING

NBA: "Have you tried tranquilizing them?"

FRED: "We've had quite a bit of experience with tranquilizing buffalo, but we've never been able to get it to work on a female. We've had a lot of help from veterinarians but they haven't had any better luck. They can't explain it, either. But we've always had good luck tranquilizing bulls.

"We had a cow that we tried to put down a couple of times to repair a prolapse. Never could do it. We'd get her down and as soon as we tried to work on her, she'd come awake. Finally, after we had enough stuff in her to kill her twice, we got her groggy enough to tie her down and work on her. She never did go out, though. All that tranquilizer didn't hurt her. She's out there running around someplace now.

"If you were to go out and try to tranquilize a good healthy cow when she's quiet, maybe it would work. I don't know. We've never tried. It's not the tranquilizer, either, because we've not only overdosed them, but we gave that one three different kinds and the best we could do was get her groggy.

"We don't shoot them with tranquilizer when they're riled—it doesn't work. Anytime we've tried to tranquilize, it's because there's something wrong with them; maybe that's part of it."

NBA: "Any advice for the beginner?"

FRED: "I'd tell him to start with calves. Then he could do like we did with our corrals. If he made mistakes, he'd find that out working his calves and he'd be able to correct his mistakes before they got too big to handle. It probably wouldn't cost too much to correct them.

"I'd say don't try to start with buffalo unless you have a good fence. Be prepared before you start or you're likely to have problems. Buffalo are not cattle, and if you don't know that when you start, you shouldn't buy buffalo."

LIKES HIS JOB

With all the hard work and challenges—knowing Fred, probably *because* it's a tough job—he enjoys his work. "I'd like to work as long as I can keep my end of the work up. I get a lot of pleasure out of working out there with the buffalo. That isn't the reason I started, but it's one of the reasons I've stayed and want to keep on working as long as they'll let me. I think it's good to stay halfway trim. I don't think I'd like to lay around and get fat. This is hard work, but it's a good job and I've come up the ladder. I've learned how to keep my crew going and when I can't keep up with them young fellers, I sit down and rest and let them do it for awhile. My experience, and the direction I give them, makes up for their youth and energy.

"Working with good people goes a long way, too."

22
Behind
the Scenes
of Public Herds

YELLOWSTONE BISON MANAGEMENT POLICIES

by Mary Meagher, Ph.D.
Research Biologist
National Park Service
Yellowstone National Park

(Edited transcript of Dr. Meagher's talk to the NBA meeting on September 10, 1982, at Cody, WY.)

Yellowstone Park represents one end of the spectrum of National Park Service areas because of both legal mandates and because of its large size and its age. It was the first national park. It does not have boundary fences. Actually, it predates the establishment of the adjacent states of Idaho, Montana and Wyoming.

This is one place where wild bison survived. Yes, some bison were brought in. But the wild bison survived and they did thrive. And as we trace Yellowstone's history, I think we can trace the evolution, if you will, of a philosophy.

When the park was established, there was no thought given to wildlife or even the protection of wildlife. There was a prohibition against wanton destruction and I don't know that anyone tried to define "wanton destruction." Camp hunting was possible, illegal market hunting was rampant, and the bison were poached very nearly to extermination by the turn of the century.

When bison were brought in—because it was feared that bison everywhere would become extinct, and in particular that the remnant wild bison of Yellowstone would undoubtedly die out—a ranching operation came into being. And some of the subsequent programs in Yellowstone were an understandable outgrowth of that operation and of the state of the knowledge of the times.

So from the time that Plains bison were brought in until the early

1950's, there was some form of active management in terms of winter feeding. Let's call it manipulative management. And again as an understandable outgrowth of the knowledge and of artificially maintaining large numbers of bison in one part of the park, a ranching concept was a part of that time.

As an outgrowth of that time, it was believed that the park would need to continue to crop bison, and there were sporadic croppings of bison in some locations in the park until the mid-1960's. There has been no human interference with the populations of bison within the park since that time.

In that period, we have been studying three different population units, in a sense three different populations, because of the environmental factors that impinge on them. Those environmental factors are different because the park is very large. And we had been watching one population, that was essentially stable, in some form of dynamic equilibrium.

Yellowstone's philosophy and legal mandates both require management to the greatest possible extent as a natural system, as a system that predates biologists, park superintendents, and indeed modern man. Animals do die; that is the form of cropping, what we call winter kill. I have heard it termed starvation. Actually, under-nutrition is probably more accurate, because Yellowstone is a very harsh environment.

We have many dependent animals that are supported by dead bison, dead elk, whatever. Whether those animals are eagles, whether they are grizzly bears—which have vanished from most of the lower 48 states— whether they are coyotes or whether they are something that would not occur to most of us without some close looking, such as bluebirds.

In the spring, the insectivorous birds can have a very difficult time if it is a late winter. And the invertebrate fauna [worms, insects, etc.] that thrive in winter-killed carcasses can be a major food source for such things as spring birds.

In other words, nothing goes to waste. Not by human standards—I think we are very used to looking at things in terms of human use. That is both understandable and appropriate in most places. But having a park such as Yellowstone, where we do not interfere, gives us a reference point as well as an opportunity for these other species to survive and to live as their kind has lived since the ice age.

Now, most national parks do not have the size of Yellowstone nor the degree of intactness. I don't mean to imply that Yellowstone is entirely intact from modern man's influence. In many ways, the more we study the park, the more amazed we are at the degree of intactness of the Yellowstone ecosystems.

Most other areas do not have that luxury. They were established later, they are often smaller. Wind Cave National Park comes to mind, for instance, established primarily to protect caverns. The thought of what a large herd of bison might need in terms of habitat was not part of the establishment of that park. So Wind Cave has boundary fences and different legal structure.

Wind Cave has a regular cropping program. This is true to some degree on some time schedule of every other national park as far as bison are concerned. Yellowstone is the sole exception.

I think with that as a brief overview, I will mention that the legal objectives and mandates for national wildlife refuges are not the same as for national parks. Parks try hardest, within their framework, to present intact a natural scene. Are there questions?

Q. Do you have any hunch whether the Yellowstone bison herd could ever crash—be entirely wiped out due to diseases or other factors?

MEAGHER: Well, we had a marked drop this winter, which I have been anticipating. We had one unit, the largest I mentioned, Mary Mountain, that has been in a marked increase phase since we ceased reductions. And then we had two very mild winters in a row and this winter was not exceptionally severe; it was slightly above average.

But with that coming at this time, with the size of the population and after two very mild winters, we did have an appreciable winter kill. In a situation such as Yellowstone, in terms of the habitat, I can't envision a crash. For one thing, the habitat is not stable. It is always fluctuating—some mild winters, some very severe. The population over time is constantly in response to that fluctuating habitat.

Q. What are the Yellowstone Park management criteria?

MEAGHER: The act of Congress that established the National Park Services defined what is a compromise for all of us—that is, the areas are both for use and for preservation—and those two things are constantly in conflict and become management decisions. Some of our decisions, with hindsight, appear to be better than others. But basically the area has not been upset by man. The natural scene prevails under that mandate.

Q. What was the object of bison research begun in the 1960's, and what was the result?

MEAGHER: I began work in Yellowstone in 1959, at which time there was a management program for bison, their four winter ranges (two are actually one population)—the Lamar, which is the northern range; Mary Mountain, which encompasses the Firehole; Hidden Valley and Pelican. And we had set management numbers of so many bison for each of these.

I tried to trace the basis. I am the researcher involved. The Park Service did not have research money. The entire Park Service budget in the early '60's was $20,000 out of Washington. There were no research proposals and there was no research staff.

I was interested in working on a Ph.D. I talked to the people for whom I worked in Yellowstone and they said, "Yes, we will give you administrative leave to go to school without pay." I paid my own costs. We bootlegged some things and toward the late '60's research as a program in the National Park Service began to have a monetary footing.

So there were no research proposals reviewed from that standpoint, there was no formal funding from the government nor from grants. It was a pieced-together kind of thing. And when it comes to a foundation where there are numbers, there was no research data. There was no range data.

There was a memo dated November, 1956, that said, "I have been out"—not myself—the one lone research biologist spread across the entire park and some other places, too, said, "I have been out and it looks to me as though there is a problem. And if I am right, then we should set these trial numbers and we should then evaluate the results."

In two years he was gone. The words "trial" and "evaluate" vanished and we had a hard and fast management program for which everyone assumed there were file drawers full of data. Does that answer your question?

Q. No. I'd like to know how much has been budgeted, how it was put together for research and how much money has actually been spent.

MEAGHER: All right. I have done almost all of the work. And I was not budgeted for, nor funded for, until I became a service research biologist in 1968. Thus, the service has spent my salary for some 14 years not only for bison but for many other things that I do in the park.

Q. This has all been "in-house?"

MEAGHER: That's right. In addition, I budget approximately $2,000 a year for aerial census work with the Piper Supercub. I keep one saddle horse and a truck and horse trailer. And the proposals are informal.

Research results are constantly subject to peer review. We are required to publish in journals and so on.

Q. I do know the Park Service has funds for research. Has Yellowstone, have you, ever applied for research monies for these ongoing research projects?

MEAGHER: Not for bison, no. We have applied for research monies for many other things. And I think it is important to realize that while I focus on bison, bison are not in a sense biologically separate from many other things. In other words, we had contract studies on climate; climatic data have done a great deal for our long-term thinking on factors that affect these large ungulate populations, as one example.

Q. You remarked about Yellowstone's intact ecology and it not being affected by human influence. I know from my own experience that even the Antarctic has had minimal disturbance and is a huge area, but still has suffered considerable disturbance. How can you say Yellowstone, which can be compared to a county in size, has not been ecologically disturbed?

MEAGHER: My comments were relative and perhaps taken too

strongly. We know there's outside influence. What we've found is that, to a surprising degree, certain of the ecosystems are very intact. Yellowstone is becoming more of a biological island. We are beginning to pick up some things such as acid rain. We have introduced timothy (grass) that has become established in many areas of the park but it does not appear to have changed the forage base in a negative way.

I think it is important to try to define those influences and what they have done and have they upset the ecosystem? There's something we need to know and recognize, but we feel we are farther ahead to try to do that, than simply look at Old Faithful on the 4th of July and assume that modern man's impact has been devastating to the degree that the system has been changed in some major way.

Q. Do you also have a plan that addresses the introduction of diseases from the outside in the animal populations?

MEAGHER: That is a possibility. And we are very concerned, for instance, that we have domestic sheep once again just about on the park boundaries. Bighorns are subject to certain domestic sheep organisms. And that is a question that is asked with anything we deal with, 'is this caused by modern man?' 'What has been modern man's influence?'—before deciding what the problem might be. Is it a natural problem, and if we can define modern man's influence, are we sophisticated enough to do anything about it or would we do better to back off and leave it alone?

Q. How do you monitor your herds? Do you keep track of how many calves are one year old or how many bulls are real old? Do you keep track of how many are dying and their age and so forth?

MEAGHER: Yes, we do know. In the winter kill this year in the Mary Mountain area, roughly half would have been calves coming yearlings. And that is not unexpected. We see the same pattern in elk.

I have a rough idea of age classes but those are strictly visual field classifications. And obviously they are subject to what any field classification of that kind would be. In terms of total numbers and calf percentages, I fly those areas about four times a year. But the numbers are not necessarily comparable because of the rut, because of shifts.

My base line count is normally in February. And what I see at other times in the year helps shape my thinking. It may tell me some things that are going on. But I don't view a November count, for instance, as a base line count.

We know we have a sex ratio of roughly 60 cows to 40 bulls. A few animals are in the 20-year-old bracket. Visually, I can do pretty well on calves coming yearlings and on bulls up through four, and then a subadult, an adult and an aged category. That is not possible on cows.

Also, those are sample counts. There is no way I can survey the entire 2,000 population. So I try to classify large enough groups that I feel the percentages will be valid for that time.

Q. What would you do if your calf population started dropping off?

MEAGHER: I wouldn't worry. In a natural system, philosophically, it requires more energy from the system for animals to reproduce than to be balanced by mortality. If young animals die and if calves are not born, if the reproduction rate is lower, then your system is not balanced by as much mortality in your adult animals. I see no reason to think that there would be any cessation in calf production. These are fluctuations according to annual fluctuations in the environment.

Also, we have essentially three population units, and the factors that impinge on those units are different. The environment in Pelican is extremely harsh, and invariably the calf percentages there are lower. But Pelican Valley is where the wild bison survived.

Q. What influence have predators had on the population?

MEAGHER: Very little. For one thing, we never had the sizeable population, as far as we can tell historically, that the number of ungulates might lead one to believe. We might have had 40 or 50 wolves at one time, and that would be three to five packs of four to six animals, let's say, and the rest would be singles. Wolves, we would expect, would have more influence on numbers. Also, they would be more apt, whatever influence they exerted, to take more readily-available prey, which would be elk.

So I would see bison as being important at times to the survival of individual large predators, or even a population such as grizzly bears in the spring. But the predators in turn would have minimal influence on bison population. I am not saying "none" and I can't define exactly what that is, but I would expect limited influence on distribution.

Q. How many bison in the park?

MEAGHER: I estimate around 2,000. I normally fly a base line count in February. When a population is up near its ecological carrying capacity, winter kill will be much heavier than at lower population level.

Q. When was brucellosis first diagnosed in the park elk?

MEAGHER: 1916.

Q. And the last one?

MEAGHER: With the reduction of 1965–66. And because I have no better data, I just assume that the percentage is the same. It runs pretty high in the bison.

Q. Are the cattle ranchers in the area of the park still complaining about this?

MEAGHER: Yes. We still do not know of a case that has ever been traced to the park.

Q. As I recall, the diagnosis was not based on blood tests but on abortion. Is that correct?

MEAGHER: No. The diagnosis was blood tests. And the percentages are all based on blood tests.

282

Buffalo in pens in New York.

Crates, buggies and crowd near pens at 1907 unloading.

283

Q. What's the Yellowstone budget for last year and what amount was spent on biological research?

MEAGHER: Our overall budget is between $9–10 million. The biggest chunk of that is maintenance. $10 million has been our budget for many years, but inflation has eroded that.

We now have two in-house research biologists—myself and a plant ecologist. We have a secretary. We have our base funding. And I pay the salary of one who does not work for me. So my overall budget to work with is probably somewhere around $100,000. And then I do have some ways for shopping for contracts that do not come out of either mine or the park's budget. There is also a lot of research that the Service does not fund, but encourages. We have approximately 150 cooperating researchers in a given season, from geologists on.

Q. How many numbers are you talking with winter kill?

MEAGHER: For this last winter, maybe 500. And the count of bison in November was 2,500. I think our dynamic equilibrium, population over time, will be 2,000 or a bit less. I would not expect to see 2,500. That was the result of a number of influences.

THE WICHITA MOUNTAINS BUFFALO HERD

In 1905, William T. Hornaday and others organized the American Bison Society. They demanded that the buffalo be given care and protection. Through the efforts of the American Bison Society and the New York Zoological Society, an offer was made to donate fifteen buffalo to the government if Congress would appropriate sufficient funds to fence an area in Wichita Forest and Game Preserve in Oklahoma. Congress set aside $15,000 for this purpose and the fence was built. On October 11, 1907, fifteen of the finest buffalo from the New York Zoological Park were shipped by rail to Oklahoma. Seven days later, these six bulls and nine cows had safely returned to the prairies.

There was great excitement in the town of Cache when the train arrived with the buffalo. The Commanche Chief Quanah Parker was among the crowd that came to the station. People from miles around flocked to the Wichita Mountains to see the buffalo.

The herd was originally kept in a pasture of 8,000 acres. Barns and feed were provided for the buffalo, and "brush arbors" were built to protect them from the summer sun. The herd prospered and grew in number.

In 1935, the Wichita Forest and Game Preserve became the Wichita Mountains Wildlife Refuge, and the entire Refuge was fenced. This allowed the buffalo to roam freely over all 59,000 acres of the Refuge. Several lakes and stock tanks were constructed at strategic locations around the Refuge. Today every effort is made to maintain the herd in as natural an environment as possible. The animals do not receive hay or concentrated feeds, but thrive on the lush native grasses.

Herd numbers are kept at about 600, and are adjusted annually to remain in balance with range conditions, and to co-exist with longhorn cattle, elk, white-tailed deer, and the other animals that utilize portions of the Refuge's forage. Approximately 100 animals are sold at live auction at the Refuge's sale arena each year. All animals are sold to the highest bidder. Health certificates for interstate shipment required by federal authorities, covering general health and certifying the animals to be free of tuberculosis and brucellosis, are provided by the Refuge.

Refuge cowboys must be ever watchful for charging animals.

The Refuge herd enjoys a modified accredited brucellosis free status, and—as of 1982—heifers are not vaccinated against this disease. Prospective purchasers are advised to be familiar with their State's laws regarding vaccination if they plan to import buffalo from Oklahoma. To avoid transmitting infectious animal diseases, all vehicles that have previously been used to haul livestock must be cleaned prior to entering the auction area. Food and drinks are not available at the Refuge.

The 1982 auction produced the average price per animal of $325.29. The 1978 auction recorded the highest price ever paid for Wichita Mountains buffalo, when $1,200 was paid for a cow/calf pair. About 70 prospective purchasers register each year, about half of this number being successful bidders for buffalo.

Prospective bidders, and others seeking further information about the Wichita Mountains buffalo herd or a pre-sale notice, should write to the Refuge Manager, Wichita Mountains Wildlife Refuge, Route 1, Box 448, Indiahoma, OK 73552.

WHERE THE SPEARHEAD BUFFALO ROAM

by J. Hall Sherwood

Buffalo on the range are part of our Western heritage that still resists taming, that lives and breathes and brings us to the brink of danger. There is a fascination in watching them that is difficult to explain, but apparent to those who have been in close touch with herds of bison and hordes of people.

Such a man is Robert Johnstone, superintendent of the state of Wyoming herds branded with the spearhead symbol. One herd is at Thermopolis. Others are in various areas throughout Wyoming. Johnstone came to Thermopolis when he was three, has worked around ranches all his life and has a fondness for livestock. He has been in charge of the Thermopolis herd since 1963.

The herd is kept under 40 to prevent overgrazing. Credit for their superb condition belongs to Dr. Dave Asay of Worland. A grain ration with added vitamins is fed. Cows are allowed privacy for calving and rarely have problems. (One young bull from this herd was recently used as a model for the cover on a publication concerning buffalo.)

Tough and hardy through the worst of winters, not inclined to over-eat, producers of meat that is nutritious and virtually fat-free, buffalo would seem easy to keep and care for. However, as many have learned, they have their paradoxes in health and handling. They are difficult to tranquilize, being susceptible to overdose. They are unpredictable, not herded as easily as cattle, and can be taken ill quickly if under stress conditions. Deep and clear comes their call to migration. Ordinary fences do not hold them at these times.

How do you explain the ironic story of the cow who was in the habit of escaping each year at calving time? Johnstone related that she insisted on returning to a certain spot to calve, about seven miles from town. After the old style wire fences were replaced by the new fence she could not get out. That year her calf was a breech birth and the cow died.

Handling buffalo takes know-how. They are easily choked by roping. They can hit up to 40 miles an hour on the run, so can wear out horses quickly and give neighbors fits as they cross boundaries. The Thermopolis herd is always moved with a pickup, dropping off rations and the buffalo trailing in to get it.

Johnstone showed a holding chute he has modified, reinforcing it with heavy wood along the sides. He has discovered that buffalo panic at the sound of clanging metal, seem more likely to calm down at the feel of wood.

Only one type of gnat has been noticed around the herd. They seem immune to the heelflies which pester cattle. Brucellosis has been kept out of these Wyoming herds. Inspections of new stock are routine. Studies are being undertaken in some states to discover the reason for buffalo's natural resistance to cancer.

On incidents involving injury to tourists, Johnstone stated that he believes "overconfidence, drinking too much and taking a dare are the main factors."

About once a season at the State Park in Thermopolis there is an episode. One spring it was a father taking a picture and his little girl getting too close. The cow charged. Injuries were not fatal. Johnstone always advises people who are taking pictures to stay in their cars. He also pointed out that irresponsible journalism "could get some one killed." He referred to the paperback *Wyoming*, with statements about a buffalo's eyesight being so poor that a man could run up behind one and hit him with his hat.

In earlier years, when fences were not as strong, he told about two bulls who escaped the pasture and carried their fight clear through Thermopolis before being rounded up. Even with the heavier fencing, occasionally an animal will get out. Then Johnstone calls on others to ride with him.

A BRIEF HISTORY OF WIND CAVE NATIONAL PARK

Wind Cave National Park is closely tied to the history of the cave from which the Park later acquired its name. The early private developers of the cave, two families named McDonald and Stabler, became involved in the exploration of Wind Cave in the 1890's. Many turn of the century tourists came to the area by tallyho stage to see the sights of "Wonderful Wind Cave."

Around the turn of the century the federal government became interested in the area, verified that there were no legitimate mining claims on the cave or surrounding area, and decided to set aside the land as a National Park. On January 9, 1903, by decree of President Roosevelt, 10,500 acres including "Wonderful Wind Cave" became Wind Cave National Park.

In 1912 an area adjoining the original 10,500-acre park was established as a National Game Preserve. Buffalo were introduced to the new preserve under the direction of the National Bison Society in 1913 with a shipment of 13 head from the New York Zoological Gardens.

In 1916 six head of buffalo from Yellowstone National Park were introduced to the Game Preserve adjoining Wind Cave National Park. We know not where any of these animals originated, but we can assume the Yellowstone buffalo came from the Buffalo Jones herd. Since 1916 there have been no other outside introductions to the Wind Cave herd.

In 1935 the National Game Preserve merged with, and became a part of, Wind Cave National Park. To the north of the land that was the National Game Preserve was a section of land that was established in 1920 as a State Park National Game Sanctuary. This parcel of land was owned by the State of South Dakota, but was administered by the federal government as a National Game Sanctuary. It sat between Custer State Park and Wind Cave National Park and, though owned by the state, was never a part of the State Park. Game Preserve buffalo roamed freely on that state land. The Game Sanctuary was annexed and became a part of Wind Cave National Park in 1946.

Wind Cave National Park, which was formed to preserve a cave of national significance, is now best known as a wildlife refuge with its herd of approximately 350 buffalo the prime tourist attraction.

BUFFALO HELP SUPPORT CUSTER STATE PARK

Nestled in the southern Black Hills in western South Dakota is a unique 73,000-acre area where the prairie meets the hills. This area, originally called Custer State Forest, officially became Custer State Park on July 1, 1919. Custer State Park provides a natural habitat for 1,000 buffalo, 500 elk, 150 white-tail deer, 120 bighorn sheep, 120 pronghorn antelope and countless small native animals. It came into existence largely through the efforts of one man, Governor Peter Norbeck.

The general philosophy of CSP buffalo management is to maintain a healthy, disease-free status with maximum production and minimum disturbance of their environment. This is accomplished with minimum handling and stress to the animals, which constitute one of the nation's largest public herds.

Buffalo were first introduced in the fall of 1914 with 25 animals from the Scotty Philip herd. In 1951, 60 buffalo from the Pine Ridge Sioux Indian Reservation were added. There is no record of any other introductions to the Custer State Park herd.

Though records don't indicate the number of buffalo maintained in the park, in its early years, there are records of cropping and harvest of surplus buffalo as early as the 1920's. Through the history of the park, surplus animals were periodically removed. This was generally done in conjunction with specific events, such as area stock shows, celebrations, and the like to effect live animal sales, and through the park's slaughtering facility for meat sales.

For approximately 40 years, little was done to keep the numbers down and no studies were conducted to establish range carrying capacities.

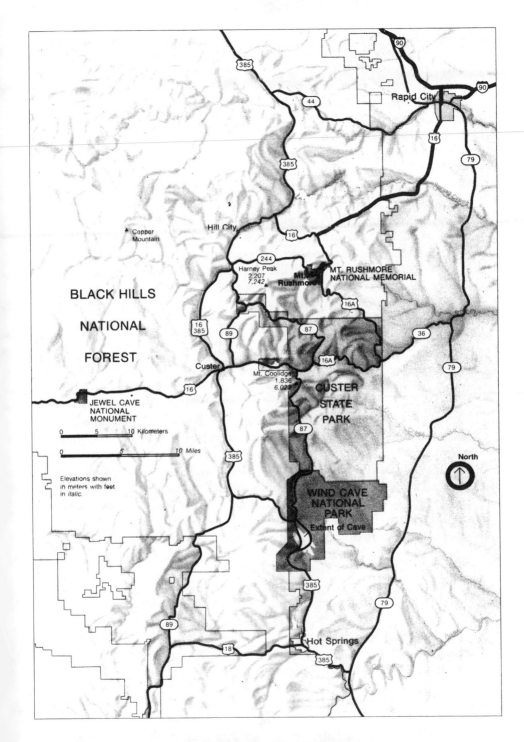

Map of Wind Cave National Park area.

Records indicate there was a fear of overstocking and the need to crop more aggressively as early as the 1940's. By the late '50's—even with periodic cropping—the herd had grown to around 2,600 head and damage to the environment was visible.

Around 1960, annual buffalo sales began and in 1966, with brucellosis eradication in the park accomplished, regular organized herd reduction sales of live animals began.

Range studies were conducted in the 1970's to determine optimum base breeding herds for the buffalo and other big game animals in the park. It was determined the park could carry from 850 to 950 buffalo. The reduction of the herd greatly improved range conditions, and insured a steady calf crop increase. The calf crop now averages 90% to 92%. Prior to herd reduction, the calf crop was at best 80% and at worst 60%.

Although the seven-foot border fence retains most of the buffalo, some do find cattle guards no real challenge to cross. Interior fences divide the range into large areas, giving park personnel a chance to manage for range and streamside improvement. Buffalo were doing a lot of damage to streamside vegetation and are now fenced out of these areas during critical growing seasons.

The adult bulls generally range throughout the park, often seeking wet, lush meadows with a nearby stream or spring. The big bulls often form bachelor groups of two to eight feeding areas separate from the cow/calf herd.

The base herd is maintained with a 10:1 cow/bull ratio. Approximately ⅔ of the animals are at breeding age, the remaining ⅓ are calves, yearlings, and two-year-olds. All calves are year branded and cows are sold at age 10. A few large, old bulls (10 years or more) are maintained for the annual trophy hunt where 10 bulls are offered in December of each year. The park tries to maintain a balance by year with approximately 70–80 in each age class up to 10 years of age. This helps to keep the number of ten-year-old cows offered each year fairly stable and keeps a good age balance among the breeding bulls.

The buffalo are rounded up the first week of October. Bulls three and older are not run through the live sale. The two-year-olds are usually still with the cow/calf herds at this time, which enables park management to gather the two-year-olds and put them in a separate pen to pick out the better bulls for sale and for addition to the breeding herd. The lower-quality bulls, with broken horns or other less desirable features, are selected for slaughter to meet the needs of the private leased concessions and lodges within the park.

During the roundup, the buffalo are run through a selection, testing, branding, ear tagging, and vaccination process as quickly as possible to reduce stress on the animals. The annual roundup enjoys the company and assistance of volunteer help from all over the country.

One thousand park buffalo have 50 acres per head to roam on. *Bruns*

The corrals are set up to divert calves from adults in a "Y" formation. The calves are worked as fast as 90 head per hour. The roundup begins on a Monday morning and usually all the animals are corraled by noon. In 2½ days, all 1,400 have been worked and the base breeding herd is roaming freely on the range once again, with only the selected sale animals remaining in the corrals.

Following the roundup, the sale animals are held in the corrals and are fed range cake and hay and allowed to graze on a half-section. Some are turned out into a two-square-mile pasture. All sale animals are blood-tested for brucellosis during the selection process and then re-tested 30 days later when they also get a TB test. The sale follows receipt of the results of the second test and is conducted each year on the third Saturday of November.

Large crowds of buyers and spectators gather for the annual Custer State Park buffalo sale.

Winter poses no threat to the buffalo of Custer State Park as the winters are relatively mild and occasional deep snows give them a chance to use their broad noses for shovels in their instinctive feeding behavior. The buffalo graze on native grasses, selecting the lushest grasses in the summer and leaning toward coarser grasses and more roughage in winter.

Salt and minerals are located strategically through their grazing areas, as are natural streams, springs and man-dug wells. The buffalo will occasionally pass by one water source in preference to another, supposedly

Auctioneer's stand
at the sale.

more desirable, source. They frequently stay in one general area for several days; but when the urge strikes, will move as much as four miles a day. Only under the most extreme weather conditions (which occur very seldom in the hills) are they supplement fed, mostly range cake which helps to maintain their body energy in severe weather. The calves released to the range during the selection process generally reunite with the cows and winter with them in the large herds.

Range cake is also used as a management tool to help keep the buffalo gentled and maneuverable. Since the herd is seldom handled and basically wild, this means of management only works when they want it to. During the calving and rut seasons, even a tasty treat fails to tempt them.

In the spring, natural weaning behavior will begin in March with the early calving cows. The first calf is usually born around the beginning of April. By this time, 10 to 25 cows may have already separated from the main herd to seek out a high mountain meadow as their maternity ward. After about two weeks of seclusion, the cows return to the main herd with their new calves. Here they congregate in groups of 300 to 600 and graze the open range till around July fourth when breeding season begins and the bulls begin mingling with the cows, calves, yearlings, and two-year-old bulls and heifers.

The breeding behavior is generally intense for about a month, with a few injured bulls leaving early to tend to wounds. Some of the younger bulls, and occasionally an older one, will stay with the cow/calf herd as late as September. These bulls are credited with breeding the cows that may have not settled during the rut, or yearlings that cycle later in the summer.

In recent years, Custer State Park has had an increasing number of two-year-old heifers calving. This phenomenon is attributed to the general good health of the herd and excellent range conditions.

By the October roundup, the older bulls are again scattered throughout the park and nobody tries to round them up.

Although the park herd produces more than 500 calves annually, only one or two cows experience calving problems each year. This is fortunate for park personnel as buffalo cows don't welcome help from man.

The herd is maintained in a healthy state through management and natural influences, with the majority of the bulls under 14 and the cows 10 years old or less.

Calves born late in the summer are branded with a smaller brand to reduce stress. During roundup and holding periods, the corrals are watered to minimize dust and related problems.

The buffalo in Custer State Park exhibit natural behavior as there is little restriction to their grazing areas and they are interrupted from their usual routine for only a few days in October.

With over 90% calving success, rare mortality, occasional twins, yearling breeding, good conformation, and a disease-free herd, the management of Custer State Park has many reasons to be proud of its accomplishments.

CSP is said to be the only self-supporting state park. Buffalo revenues form a large percentage of this support.

HOW WE RUN THE NATIONAL BISON RANGE

by John Malcolm
NBR Manager

(Editor's Note: The following is an edited transcription of Mr. Malcolm's talk to the NBA convention Sept. 10, 1982, Cody, WY.)

The U.S. Fish and Wildlife Service operates a system of approximately 425 National Wildlife Refuges across the nation and that includes the recent addition of several large new refuges up in Alaska. This system now contains about 8 million acres.

Most of the refuges were established and are managed mainly for migratory waterfowl, but there are a number established for big game species. Of the big refuges, four were set up by special Acts of the Congress for preservation of the American bison. These four are the National Bison Range in Montana, the Wichita Mountains Wildlife Refuge in Oklahoma, the Fort Niobrara Wildlife Refuge in Nebraska and Sully Hill in North Dakota.

National Bison Range corrals in the Flathead Valley of Montana.

The two range herds are rotated on eight fenced-off pastures within the National Bison Range.

I'd like to briefly touch on some of the policies that apply to all four of these federal refuges that run buffalo. One of the policies of the sysem is that buffalo will not be introduced into any other refuges other than these that specifically have bison in the establishment legislation. The objectives of the areas are to assure a nucleus breeding population of the species; to provide appropriate viewing opportunities for public enjoyment; and to support scientific study which is feasible within the scope of managing representative bison herds.

In the area of herd composition, the basic policy is that an even 50-50 sex ratio will be maintained. And age composition of the herds is approximately what occurs under natural conditions. In disease prevention and control, managers of the federal refuges are to coordinate with state and federal veterinarians and comply with all rules and regulations. The managers are also to develop programs to provide for periodic inspection and testing of the herds to insure that they remain in good health and do not carry infections.

Disposal is normally made only during the regularly scheduled disposal period, except in emergencies. This position is effected by sales—except that surplus animals may be donated to zoos and other tax-supported institutions. Live sales are encouraged. Meat sales are the last resort for disposal of excess bison.

Now I'd like to give you an idea of the management and operation of the National Bison Range. The Bison Range is located in Flathead Valley in northwestern Montana. It's about 45 miles north of Missoula, contains 18,500 acres and is actually kind of a small mountain range that stands by itself in the valley there. The differences in elevation from the low point around the outside of the range to the high point in the center is over 2,000 feet and the terrain is fairly rugged.

Most of the area is open bunch grass habitat, but there are some Douglas firs and ponderosa pine in the high portion of the range. We run approximately 325 buffalo in the base herd at the Bison Range and, with the calf crop every year, it will get up over 400 animals.

The buffalo there are run in two main herds. We have eight different pastures that are fenced off on the range. Each herd has four pastures. And each herd is rotated on a different rotation grazing schedule every three months among its various pastures. Under this setup, any one of the range pastures is used during the critical growing season only one year out of four. Also, each pasture gets a full 18 months' deferment.

In the annual cropping program on the bison, there's approximately 80 head sold each year on a sealed bid basis. The bids are generally put out by mid-August and are due back in our office the first week in September. The animals are rounded up the last week in September and run through the corrals. Animals to be sold or disposed of are picked out. We vaccinate all animals for lepto and heifer calves are vaccinated for brucellosis. All the calves get a year brand so we can tell their ages in the future. In 1981, we started a program of bloodtesting all the calves for a study being conducted by Dr. Stormont of Stormont Labs in California.

We maintain about 55% cow to 45% bull sex ratio in our herd and the age spread, after the roundup each year, is about 25% calves, 15% yearlings, 35% two- to five-year-olds, 20% six- to ten-year-olds and only about 5% 10 years and older.

With the 55/45 cow/bull ratio on the National Bison Range we run a calf crop that averages close to 90%. With our sex and age ratios we generally have around 100 breeding age cows.

A new visitor's center at the range was completed in 1982. It has displays and exhibits on the history of the buffalo and on the management of the National Bison Range. In addition, we have two tour routes open to the public and a 20-acre irrigated pasture where we keep a few on hand close to headquarters, so visitors are just about guaranteed seeing at least a few animals. We run about 100,000 visitors a year at the Bison Range.

The original animals came mostly from a herd at Kalispell which previously had come from the Pablo-Allard herd. With the original bunch there was also one from the Goodnight herd of Texas and three from the Blue Mountain Forest Association herd in New Hampshire. Then there have been some animals brought in from other sources over the years.

CANADIAN WILDLIFE SERVICE REHABILITATES WOOD BISON

by H. W. Reynolds
Canadian Wildlife Service

The wood bison (*Bison bison athabascae*) is scientifically recognized as a subspecies of North America bison. In Canada, historical distribution included most of the boreal regions of British Columbia, Alberta, Saskatchewan, Northwest Territories (NWT), and the Yukon Territory. Wood bison also ranged along the eastern slopes of the Rocky Mountains from northern Canada to Colorado in the United States.

Problems for survival of the subspecies began in 1925 when a decision was made to introduce excess Plains bison from Wainwright Park in central Alberta into Wood Buffalo National Park (WBNP) in northern Canada. From 1925 to 1928, approximately 6,673 Plains bison were released along the west side of the Slave River between the Peace River and Fort Fitzgerald. Immediate herd mixing with the estimated 1,500 resident wood bison was noted. By 1940, it was generally believed that pure wood bison no longer existed.

During an aerial survey of WBNP in 1957, N. S. Novakowski, Canadian Wildlife Service, discovered an isolated herd of bison in the northwest corner of the park. Subsequent sampling suggested they were of wood bison stock. In 1963, 18 of these animals were transferred to an area northeast of Fort Providence, NWT, to isolate them from an anthrax outbreak in bison along the Slave River. Continued concern regarding anthrax led to a transfer of 23 wood bison to an isolation area in Elk Island National Park, Alberta in 1965 to save the subspecies from extinction and to establish a source breeding herd for future transplants. The Fort Providence herd now numbers about 600 and the Elk Island herd 120.

Wood bison are officially recognized as endangered. Provincial and territorial governments share with the federal government responsibility for their rehabilitation.

The wood bison rehabilitation project commenced in 1975. Its primary objective is to establish a minimum of three free ranging and self-perpetuating populations of wood bison in areas of historic range. The second objective is to protect and preserve the gene pool by dispersing small breeding herds to zoological gardens and parks. With respect to the latter, animals are transferred according to the terms of a lease agreement whereby the federal government retains ownership of all wood bison and their progeny. Breeding stock has not been made available to private or commercial enterprises because of problems associated with record keeping and the lack of capability to meet the expected demand.

The ultimate goal of the program is the reestablishment of free ranging and captive populations of wood bison in sufficient numbers to warrant its removal from the lists of endangered species. The project will then terminate and commercial trade in wood bison will be possible.

Successful transfers have been made to five institutions: Calgary Zoo, Alberta, four calves in November, 1976; The Wildlife Reserve of Western Canada, Alberta, two calves in May, 1977 for behavioral research; Metro Toronto Zoo, Ontario, 10 adults in November, 1977; Moose Jaw Wild Animal Park, Saskatchewan, 10 adults in March, 1978; and San Diego Zoo, U.S.A., three calves in January 1979. The wood bison at Calgary Zoo, Toronto Zoo and Moose Jaw have produced offspring each year.

In summer 1978 wood bison from the disease-free herd at Elk Island National Park were released to the wild in Jasper National Park, Alberta. The experiment failed when the herd moved out of the Park. Experience gained from this attempt led to alternate methods of reintroduction. One is to hold young animals in a corral at the chosen site for a time in hope they will locate nearby when released.

Future priorities of the program are to continue with transplants both to the wild and to captivity by about 1986.

Little-Known Buffalo Benefits

VI
VI
VI
VI
VI
VI
VI
VI
VI
VI
VI
VI

23
The Exceptional Nutritional Value of Buffalo Meat

THE AMAZING BUFFALO

Why was it the Plains Indians, who lived principally on buffalo meat, were among the finest physical and mental specimens the human race has known? Their strength, health, keenness of sense and mind, physical endurance and longevity were legends in their own time.

Blood Blackfoot Chief Buffalo Child Spotted Calf Long Lance, in his autobiography *Long Lance,* relates a conversation between a Suksiseoketuk chief, Travels-Against-The-Wind, and their own chief in about 1895, when they were hunting down the last of the buffalo in the foothills of the Canadian Rockies.

Travels-Against-The-Wind warned the Blackfoot against the white man's food that would make their teeth fall out.

"Our people, like yours, never used to die until they were over a hundred years old. Now, since we started to eat the white man's food we are sick all of the time. We keep getting worse and soon it will kill us all."

The plainsmen and mountainmen who adopted Indian ways and diet were also capable of legendary feats. None, once he had tasted buffalo, willingly went back to beef, the sweet smell of which actually sickened Indians used to buffalo.

Besides being the most delicious of meats, according to those who especially enjoy buffalo, buffalo meat has some nutritional advantages. The most enthusiastic claim near miraculous powers for it. People who eat it regularly say they feel better, more vigorous, and find weight control easier because they are satisfied with less.

Science knows appallingly little about buffalo meat. While beef and pork, venison and fowl, fish and snails have been exhaustively analyzed, scientists have barely started analyzing buffalo. Results to date indicate that, compared to beef, buffalo is higher in protein and polyunsaturated fats, higher in a compound associated with fast muscle action (and buffalo, while seemingly slow and clumsy, are famed for their fast gallop and agility with which they can wheel and disembowel your horse on their wicked horns).

Is there a relationship between the enormous strength, health, keenness, endurance, and longevity of the buffalo—and the similar qualities of the Plains people who lived on their flesh?

BUFFALO (BISON) NUTRIENT CONTENTS AND OPTIMAL COOKING QUALITY

by Wayne A. Johnson, Ph.D.
Head, Nutrition-Food Service Dept.
South Dakota State University
Brookings, SD 57007

Many people have sampled a well prepared piece of buffalo meat—but the vast majority of American food consumers know little about this product. Perhaps the mystique of the animal, stimulated by historical accounts of vast herds of these majestic beasts thundering across the western prairies and hills, coupled with the knowledge that the bison nearly became extinct, has clouded the acceptance of buffalo as an exceptional source of tasty meat. Recently the animal has enjoyed a growth in numbers and popularity as more producers and consumers have become aware of its potential as an alternative meat source.

For most foods consumed by the American public, we have extensive information about the nutrient content and optimum cooking methods. Most of these foods have been subjected to careful testing and evaluation. This has not been the case for buffalo (bison) meat. Although the meat has become increasingly popular and the animal more widely produced in this country and abroad, there have been no carefully designed and documented research studies conducted on buffalo. As consumption has increased and the meat has become more available, the lack of nutritional information on buffalo meat has become a liability to the buffalo industry.

Food industry representatives, dietitians, medical doctors, and the average consumer have been clamoring for more information about the buffalo. Questions abound such as, does buffalo meat really have less fat, less cholesterol? Is it higher in protein than other meats? Are there fewer allergic reactions to buffalo meat?

The answers to these questions and others have now been provided in the results of a research study conducted at South Dakota State University in the Nutrition and Food Science Department. In a project designed by Drs. Wayne Johnson and Elizabeth Rust (deceased), and conducted with the assistance of Dorothy Deethardt and Louise Guild, information was collected on vitamin and mineral content, protein and amino acid composition, and the amounts of fats and cholesterol in various cuts of the meat. In addition, the meat was subjected to a variety of cooking procedures and techniques to determine which methods resulted in the most palatable and nutritious product.

Throughout, the National Buffalo Association and its individual members lent support and encouragement. Funding was from state and federal funds provided by the SDSU Agricultural Experiment Station and from private donors. Meat samples for the project were largely obtained from two sources, the L. R. Houck Triple U Ranch and the Custer State Park (South Dakota). Other selected samples were received from producers in New York, Missouri, Florida, and California. These animals were range fed, with the exception of the Houck animals, which were partially grain finished (90 days on grain plus grass).

Meat samples were prepared using a variety of cooking methods. Various cuts were roasted, broiled, braised, or cooked in a microwave oven. Results are identified according to cooking procedures used.

All chemical and physical testing of the meat samples was conducted according to methods approved by the Association of Official Agricultural Chemists. Consumer analysis of product palatability and overall acceptability followed standard methods commonly used for evaluation of meats and meat-based foods.

Nearly everyone who discussed buffalo meat as a food product has an opinion about the protein and fat content. Most proponents of buffalo believe that the meat is higher in protein and lower in fat than any other source of red meat. To some extent these beliefs were confirmed by our studies, since the average fat content of the edible lean portions of meat averaged only about 4% fat, compared to 12–15% for pork, 10% for lamb and 6–8% for similar cuts of beef. It is important to remember, however, that these buffalo were essentially grass fed, non-finished animals, while most published information for pork, lamb, and beef reports the composition of market finished animals. Meat samples from those buffalo whose feeding program included some grain tended to show slightly higher levels of fat. Thus, any statements about nutritive value of buffalo meat should also indicate the feeding history of those animals. Where comparisons were made in this article, we used published data for the lean cuts where possible.

Protein content of the various cuts ranged from 30–32% (on the average) for the range-grain animals, to 32–34% for samples from the range-fed group. This small difference is primarily the result of higher fat in the first group mentioned. These values are comparable to those for lean beef (31%), lamb (28%), and pork (31%).

Red meats tend to have fairly similar levels of most vitamins and minerals, and buffalo is no exception. We were, however, particularly interested in the thiamin content of buffalo because there are few food sources that are exceptionally high in this vitamin. Among the meats that do contain large amounts of thiamin are pork and venison. Buffalo meat was found to have approximately one-tenth milligram of thiamin per 100 grams of lean meat, which translates into about 0.3 milligram per ounce. These animals are nearly identical to the values for lean beef, and about one-tenth that of pork.

Probably more questions and controversies have surfaced regarding the fat and cholesterol content of buffalo meat than for any other single topic. As was mentioned earlier, the average fat content of the meat analyzed by our laboratories was low; and as a result one would expect the cholesterol values to be similarly low. Our results showed the buffalo meat cholesterol levels in the range of 150–200 milligrams of cholesterol per pound of cooked lean meat. Thus, primarily because of its low fat content, buffalo does have minimal cholesterol in comparison to other red meats as normally finished, prepared, and consumed. Based on these findings, buffalo meat, with about 10 mg cholesterol per ounce of lean tissue, could be safely used in moderation in most low-cholesterol diets.

Dr. Rust found that buffalo meat has a higher concentration of essential amino acids. It has more unsaturated fatty acids and less cholesterol. She found that buffalo meat has qualities similar to the meat found in the goose's wing. These qualities make a goose capable of long, sustained flights. The study of nutrition should be an important program, since large portions of our world are populated by undernourished people.

Most meats, especially from the so-called red meat species, tend to contain predominantly saturated and monounsaturated fats, rather than the polyunsaturated varieties. Buffalo meat follows that general pattern, with approximately 85% of the fatty acids falling into the saturated and monounsaturated categories. Approximately 10% of the fat (0.5 grams per 100 grams of meat) was in the form of polyunsaturated fats.

Mineral content of composite samples was similar to values reported for other meats. We did note a somewhat lower iron content (1.0–1.5 milligrams per 100 grams of lean meat) in comparison to beef (3.5 mg/100 grams) and pork (3.8 mg/100 grams). A possible explanation for the generally lower reported levels of iron in recent studies is that the meat has been cooked in aluminum or teflon-coated pans, or prepared by the microwave process. It appears that at least some of the iron value of foods may depend upon the cooking utensil and preparation method.

Each type of food requires slightly different cooking methods, depending on the special characteristics of that product. Even when preparing similar cuts of meat, we must know the best temperature to use, whether to cook by moist or dry heat, and what final internal temperature provides the most appealing, palatable product. For these reasons, we conducted tests in which various cuts of buffalo were cooked at several temperatures and to varied end point internal temperatures. The results of this study showed that for open pan oven broiling of rib, sirloin tip and sirloin butt roasts, the preferred process involved roasting at 275° or 300°F to a final internal temperature of 155° or 165°F (rare-medium rare). Meat cooked under the first conditions was more tender and juicy than for any other combination, while the consumer taste panels generally gave the highest scores to roasts prepared under the latter conditions. In general, as the oven temperature was increased above 300°F and the final internal temperature rose to higher than 170°F, taste panel scores for tenderness, juiciness, and texture declined. This appears to indicate a need for slow cooking of buffalo, and care to not overcook.

Several other cooking methods were tested, such as open pan braising, plastic bake in bag, aluminum foil wrap, brown bag roasting and microwave oven process. These techniques are described in more detail in a SDSU recipe bulletin. Generally these methods yielded acceptable products if care was taken to prevent overcooking the meat. Less tender cuts of meat may require longer cooking periods with moist heat.

We now have firm, well-documented evidence that can be used when discussing the nutrient content and cooking qualities of buffalo meat. Hopefully this information will serve to decrease the aura of mystery and "magical qualities" which have perhaps served as a detriment to complete acceptance of the buffalo as a viable alternative meat source. Nutritionists and physicians can answer questions about the product with greater assurance, and the average consumer or food service manager will possess guidelines for preparation of the meat in such a way as to ensure the most acceptable, taste-tempting product.

NUTRITIONAL QUALITIES OF BUFFALO MEAT

(The following combines excerpts from two scientific articles: One is "Nutritional Content of Selected Aboriginal Foods in Northeastern Colorado: Buffalo (Bison bison) and wild onion (Allium spp.)," by Elizabeth Ann Morris, Colorado State University; W. Max Witkind, U.S. Corps of Engineers, Little Rock, AR; Ralph L. Tix, CSU and Judith Jacobson, CSU, from Journal of Ethnobiology, December, 1981. We assumed you'd be more interested in the bison information, so we omitted the onion part. The other is an unpublished paper by Morris and Witkind, "A consideration of the Nutritional Benefits Derived from Bison Meat.")

To identify the nutritional benefits available from prehistoric bison hunting, tests were carried out at Colorado State University. Sample material was obtained from a 500-pound yearling buffalo bull whose diet has been natural prairie grasses. Hump, shank, loin, and the gracilis and adductor (thigh) muscles in the rump were tested. Testing of wet samples indicated that between 67 and 73% of the butchering weight consisted of water, between 4 and 9% fat and between 25 and 29% protein. The fatness of an animal would vary a good deal with season and richness

305

of pasture. The total caloric content of the samples varied from 83 cal/g in shank meat to 2.14 cal/g in loin meat. The great range reflects the variation in the amount of fat (with its high caloric content) in the individual samples.

Jerky

Long ago, in the absence of refrigeration, drying meat to make jerky was a common way to preserve meat. (The term "jerky" is reportedly derived from the Arabic *charqui* or *charqi,* meaning dried meat.) The term and the practice were brought to the New World by the Spaniards and the word was then applied to the similar local Indian preservation method.

Jerky is thin slices of meat dried in the sun for a few days to reduce the moisture content, reducing the weight by ⅔ to ¾, an important consideration for people who carried their supplies, or depended on dog travois. Drying also kept the meat from spoiling. Only moisture is eliminated by drying; the fat, protein, minerals, and trace elements remain.

Pemmican is an Algonkian word for a mixture of dried berries or fruit, such as plum, chokecherries, service berries or currants. It is pounded to a paste and added to pounded dried meat, then soaked with melted fat and marrow (bone grease). The resulting product, wrapped in a rawhide bundle called a parfleche, keeps for years.

The fresh weight of 5 1-g samples of hump, shank, loin, gracilis and adductor muscles was compared to the dry weight of these samples (see Table 1.)

Caloric Energy Content

This is also presented in Table 1 as calories per gram of meat. The increased caloric content in dry meat is proportional to the amount of moisture removed. The large range of readings reflects the high proportion of fat to protein in the loin samples. It has been suggested that the relatively low caloric content of the shank meat sample might indicate that bone or cartilage was inadvertently included.

Mineral Content

Appreciable amounts of subject minerals were found in each sample, with the exception of cobalt, which was not found in the shank, loin and adductor samples (see Table 2).

Almost all of the eight trace elements were present in greater quantities in buffalo hump meat than in the other muscles tested. It is interesting to note that hump meat was greatly prized by Indians and early settlers on the Plains. It not only tasted better; it was better for them. The hump meat contained the highest concentration of calcium (.0026% fresh, .0039% dry). *Why* is unknown. It may be that the low moisture content of the hump meat sampled is in some way related to its high calcium content. Here again it is possible that the hump sample might have included a bit of cartilage or bone inadvertently.

Buffalo Jerky Diet

Only 0.7 pound of buffalo hump jerky would satisfy the recommended daily allowances of all eight minerals; 1.2 pounds would supply these and the minimum calories, as presented in Table 3. These amounts are not excessive and the nutritional value of bison meat, fresh or jerked, is obvious.

Because bison bones are so often prevalent in local archeological sites, it is interesting to compare the amount of meat available from a bison to that from other animals. A selected sample indicates that the total weight of buffalo bulls ranges between 1,600–1,800 pounds, while buffalo cows range between 800–900 pounds. A whole bull elk weighed by the Wyoming Game and Fish came in at 529 pounds, approximately one-half the weight of a buffalo cow and one-third that of a bull. Deer and antelope are even smaller.

Based on the above scientific articles, a diet featuring buffalo meat, flavored with onions, would supply a high proportion of our basic nutritional requirements.

Table 1.

Compositional analysis of Bison meat per one gram sample. Figures are rounded to nearest whole number.

Bison meat Sample	Fresh and Dry Weights	Moisture	Fat	Protein	Calories Per Gram
Hump					
Fresh	1.00 gm.	67	05	25	1.38
Dry	.33 gm.	00	16	83	4.62
Shank					
Fresh	1.00 gm.	71	04	29	0.83
Dry	.29 gm.	00	11	84	2.43
Loin					
Fresh	1.00 gm.	73	09	28	2.14
Dry	.27 gm.	00	23	75	5.77
Gracilis					
Fresh	1.00 gm.	67	05	27	1.59
Dry	.33 gm.	00	17	88	5.17
Adductor					
Fresh	1.00 gm.	69	05	27	1.68
Dry	.31 gm.	00	16	85	5.21

Table 2.

Mineral elements in Bison meat samples. Quantitative aspects of mineral elements are expressed as percent figures.

Bison Meat Samples	Ca	P	Na	K	Mg	Zn	Cu	Co
Hump								
Fresh	.0026	.3990	.0765	.3135	.0170	.0025	.0006	.0012
Dry	.0039	1.3368	.1150	.4710	.0260	.0035	.0009	.0019
Shank								
Fresh	.0020	.2565	.0555	.2310	.0150	.0050	.0002	.0000
Dry	.0029	.7500	.0785	.3255	.0210	.0070	.0004	.0000
Loin								
Fresh	.0015	.6540	.0370	.2670	.0195	.0023	.0003	.0000
Dry	.0020	1.7610	.0505	.3660	.0273	.0031	.0004	.0000
Gracilis								
Fresh	.0014	.3885	.0375	.2730	.0170	.0027	.0003	.0011
Dry	.0021	1.1020	.0560	.4065	.0255	.0031	.0004	.0016
Adductor								
Fresh	.0010	.3805	.0310	.2505	.0175	.0020	.0004	.0000
Dry	.0014	1.1770	.0455	.3630	.0255	.0030	.0006	.0000

Table 3.

Determination of how much bison hump jerky eaten per day would satisfy some of the recommended dietary allowance requirements for humans over 1 year of age. Figures are rounded to nearest tenth.

	Minimum daily Requirement Range	Grams of Jerky	Pounds of Jerky
Calcium	800–1200 mg.	205.1–307.7	.5–.7 lb.
Phosphorus	800–1200 mg.	less than 1 gm.	less than .1 lb.
Sodium	325–3300 mg.	2.8–28.7	less than .1 lb.
Potassium	550–5625 mg.	1.2–11.9	less than .1 lb.
Magnesium	150–450 mg.	5.8–17.3	less than .1 lb.
Zinc	10–25 mg.	2.9–7.1	less than .1 lb.
Copper	1.0–2.5 mg	1.1–2.8	less than .1 lb.
Cobalt	?	?	?
Calories	2400	519 gm.	1.2 lbs.

TASTE PANEL – FLAVOR, JUICINESS, AND SHEAR FORCE COMPARED TO BEEF STEERS

	Flavor[a]	Juiciness[a]	Shear force[b]
Angus steer (control)	5.1	5.6	5.19
Galloway steer	4.7	3.6	5.45
Buffalo steer	4.4	5.7	10.99

[a]The higher number was more desirable.
[b]One-half-inch core; lb. per sq. inch.

Editor's Note: Naturally we question the high shear force figure for the buffalo meat. This means it was twice as hard to chew as the beef. We do not know how the meats were cooked and can only guess that all three samples were cooked with scientific impartiality and exactly the same. In such a case, the buffalo meat would be overcooked and thus tough. Also, a 5-year-old animal is past his prime and should have been butchered much earlier.

TECHNICAL REPORT NO. 302
CARCASS CHARACTERISTICS OF A
BISON STEER (BISON BISON)
ABSTRACT

Investigators from the Animal Science Department, Fisheries and Wildlife Department, and the Natural Resource Ecology Laboratory of Colorado State University took various biological and "commercial" measurements of a *Bison bison* steer slaughtered for meat. Results are tabulated and presented without analysis. Taste panel comparison is presented. This animal was a five-year-old bison steer whose ration, unaltered for at least 180 days, consisted of:

    6 to 6.5 pounds of 16% protein pellets per day

    6% crude protein crested wheatgrass hay ad libitum and native short-grass prairie ad libitum

SUMMARY OF DATA COLLECTED FROM BUFFALO STEER BELONGING TO NREL

Slaughtered November 7, 1975

1. Dock weight	= 1051.0 lb.
2. Pluck weight (without heart)	= 20.0 lb.
Heart weight	= 4.5 lb.
3. Liver weight (without bile)	= 9.0 lb.
Bile weight	= 1.5 lb.
4. Full gastrointestinal tract weight	= 159.0 lb.
4 stomachs	= 92.5 lb.
Remainder of gut	= 67.0 lb.

5. Spleen = 3.0 lb.
6. Head weight = 61.0 lb.
7. 4 shanks = 19.0 lb.
8. Hide = 54.5 lb.

Carcass weight:
  Right side = 316 lb. (hot)
  Left side = 322 lb. (hot)

Cut up data collected on November 14, 1975:

Left side (cold)
  Hindquarter = 134 lb.
  Forequarter = <u>167 lb.</u>
  TOTAL   301 lb.

Right side (cold)
  Hindquarter = 146 lb.
  Forequarter = <u>163</u>   lb.
         309 lb.

Right side cut up data (bone in)

Round, untrimmed	= 60 lb.
Rump & loin, untrimmed	= 51 lb.
Rib, untrimmed	= 20 lb.
Chuck, untrimmed	= 100 lb.

Wrapped meat, total for both
  sides   =502 lb.
Hamburger   =117 lb.
Retail cuts   =385 lb.
Kidneys   = 2 lb.
together (not included in 502 lb. total)

Separate fat, lean, and bone of rib steak:

Rib from:	Total wt.	Fat wt.	Bone wt.	Total Lean wt.	Rib eye
Left side	147.29 g	58.04 g	30.68 g	58.57 g	38.93 g
% of total		39.41%	20.83%	39.76%	26.31%
Right side	113.40 g	44.26 g	14.58 g	54.56 g	31.92 g
% of total		39.03%	12.86%	48.11%	28.15%

Proximate analysis of rib steak:

	% Moisture	% Lipid	% Ash	% Protein
Left side adipose	7.83	87.91	0.42	3.84
Left side lean	71.46	3.00	1.10	24.44
Right side adipose	13.23	78.02	0.46	8.29
Right side lean	72.20	3.58	1.01	23.21

Carcass evaluation:

Fat thickness	Chine bone	Middle	Rib
Left side	1.3 inches	1.1 inches	0.7 inches
Right side	1.2 inches	0.9 inches	0.6 inches

Kidney knob = 12 lb.: 1.97% includes kidney

24
Put Buffalo on Your Table

WHY EAT BUFFALO MEAT?

Well, for one thing, it *tastes good!* Very good! If you didn't know what you were eating, you'd probably think that it was the best beef you ever tasted. It has been said that "it tastes like meat used to taste," or that "it tastes like beef wished it tasted." Buffalo meat is just naturally flavorful and tender. You don't need to slop up a "buffalo burger" with a lot of guck. Buffalo will make any recipe better and tastes great by itself.

And it's not greasy, which brings up another point. The flavor in most meats is in the fat, but so are the calories and the cholesterol and saturated fats! With other meats, you trim off the fat (which you paid for) and try to cook it out and drain it off. But buffalo meat is tasty without the fat.

You get your money's worth. Buffalo meat is satisfying. It's "concentrated." That's because what you pay for is mostly protein—pure, natural, wholesome protein—25 to 30% more protein than beef. It doesn't shrink all up in cooking. So pound for pound, it doesn't cost much more than "cheaper" meats.

It is *healthful.* A long time ago, the Indians and frontiersmen who lived predominantly on buffalo meat, had endurance, agility and stamina that we moderns find hard to believe—and they had few doctor bills! Some people today say they feel better with buffalo meat in their diet. There are real reasons for this. It won't upset your tummy, for one. And buffalo, due to their nature, are handled as little as possible. They spend their lives on grass, pretty much as they always did, and spend very little time in the feed lot. Their meat is not full of drugs, antibiotics, nitrites, chemicals and stuff like diethylstilbesterol (artificial female sex hormone). Buffalo are non-allergenic. People who cannot eat other meat can eat buffalo. Buffalo meat is remarkably high in actomyosin, a compound associated with fast muscle action. So far, only the wild goose wing muscle has been found to contain more actomyosin. We have barely begun to find out, scientifically, what the Indians knew all along!

Buffalo meat is great for everyone, from dieters to athletes. Buffalo are not an endangered species. The slaughtered animals are surplus bulls from commercial buffalo ranches. If this sounds just too good to be true, why don't you find out for yourself!

BURGERS ARE BETTER NOW!

A decade ago, almost the only buffalo meat that came on the market was aged bulls and barren old cows—as tasty as cardboard and about as tender. Buffalo burgers were most people's introduction to buffalo meat. The burgers then were like buckshot soaked up in vulcanized rubber sponge. Naturally, first-time eaters got turned off.

Thanks to independent buffalo ranchers (no thanks to the government), the burger situation has been largely corrected. Now the market is getting the flesh of prime young buffalo, some grain-finished. So even buffalo burgers, like steaks and roasts, are sweeter than candy and tender as butter.

BUFFALO MEAT IS A BARGAIN

Pound for pound, buffalo is about 30% higher in protein than beef. This accounts for its richer taste. You will also find this difference means smaller portions will satisfy most appetites. Though your price per pound for buffalo may be higher than a comparable cut of beef, you will find this difference more than offset, as less goes farther. And you save energy too, with the lower heat and shorter cooking process.

Cholesterol-conscious people will find buffalo, with its low fat content, especially appealing.

Weight Watchers are adopting buffalo meat as a neutral meat in many areas where it is readily available. This means dieters may eat buffalo without forfeiting an allowed beef and pork meal.

Buffalo has never been known to cause an allergic reaction in anyone. Persons whose systems react unfavorably to some food products will find they thrive on buffalo meat.

And perhaps best of all, buffalo meat is totally natural. No force feeding with questionable chemicals, no additives, just 100% pure nutrition.

Nutritive Values of Beef Versus Buffalo
A composite of cooked samples representing the whole animal.

	Water %	Protein %	Fat %	Minerals, mg/100gm					Vitamins mg/100gm
				Calcium	Phosphorus	Iron	Sodium	Potassium	Thiamin
Beef[a]	44	24	27.9	10	178	0.31	54	24	.062
Buffalo	63	35	2.8	10	174	0.43	78	37	.050

[a]Calculated from Nutritive Value of American Foods, Ag Handbook No. 456, USDA.

(Table taken from The Best of Bison *by the Agricultural Experiment Station, South Dakota State University.)*

Now that you're sold on buffalo as a food product, the next question is, "Where can we buy it?" For years, buffalo meat was available to only a select few. Many people feel this is true now, as the product seldom appears in supermarket meat cases. Buffalo meat is relatively scarce as compared to beef, but it is available if you're willing to search a little.

Most buffalo meat is available directly from the producer, generally by the quarter, half or carcass, but more and more frequently in smaller, processed bundles. Seeking out a producer, buying in bulk and having your meat processed may seem inconvenient to those accustomed to the luxuries of supermarket buying, but keep in mind that this method cuts out the middleman and retailer and will save you $$$'s.

If you haven't already found a supplier, the NBA will be happy to help you. We have 400 producers on file, from nearly every state in the USA and most provinces of Canada. These buffalo producers would be more than willing to help you satisfy your hankering for buffalo meat. NBA member producers are listed by state or region to make it easier for you to locate buffalo meat in your area. These people will also assist you in other ways if your interest goes beyond the desire to purchase buffalo meat. In areas where producers are scarce, we will help you find a supplier who air freights the meat anywhere you want it.

If you need assistance contact:

NATIONAL BUFFALO ASSOCIATION
P.O. Box 706
Custer, South Dakota 57730
(605) 673-2073

BASIC COOKING TIPS

There is no such thing as tough buffalo meat. There are only improperly instructed cooks.

Buffalo meat, though similar to beef, needs to be handled and cooked differently. You will find you can interchange buffalo meat with all your favorite beef recipes, if you follow a few basic instructions.

Individual cuts of buffalo will appear, except for color, identical to beef. Prior to cooking, buffalo meat is darker, almost brown. This is because buffalo meat does not marble like beef. The lack of marbling, or fat, is the reason buffalo must be cooked differently. It's also a good idea to trim the visible fat from around a cut of meat prior to cooking.

Fat is an insulator and heat must first penetrate this insulation before the cooking process begins. Buffalo, with its low fat content, does not need to be cooked as long with as high a temperature to get the job done. Overcooked buffalo meat will bring you the same results as overcooked beef: something nearly as palatable as shoe leather. It simply doesn't pay to rush the cooking process.

When oven broiling buffalo, move your broiler rack down a notch from where you'd place it to broil beef. Then check and turn your steak a few minutes sooner than you normally would. The result will be a steak that tastes better than the best beef you've ever eaten. The rare to medium range is best for buffalo steaks. We make no promises for well-done (overcooked) steaks.

If you normally roast beef at 325°, turn your temperature down to 275° for buffalo. Plan on the roast being done in the same time a beef roast of comparable size would be when cooked at the usual (higher) temperature. Check it, though, because it might be cooked a little sooner, or better yet, use a meat thermometer, reaching the same internal temperatures you'd use for beef. Again, rare to medium is best. And *don't cover* it. Don't even foil wrap it.

Chopped buffalo or buffalo burger will also cook faster. Buffalo burger contains little fat, so your lower temperature will also help to insure that the meat doesn't scorch. With burger the amount you see raw is going to be about the amount you get when it's cooked, so you can start with smaller patties and be assured of ample servings for all (having less fat, less grease oozes out; the meat doesn't shrink so much).

Very slow, moist heat works especially well with the less tender cuts of buffalo, such as chuck. There is nothing to compare with a buffalo chuck roast cooked all day in a slow-cooker. With slow-cooking, you don't have to worry about overcooking. Let it go till it falls apart and you won't be disappointed.

Microwave cooking of buffalo meat is a "new frontier." You may have to experiment a bit to obtain the best results. I've had some good results using microwave beef recipes with buffalo, but you have to be especially careful on your timing as a minute in the microwave can make a BIG difference. Using a lower setting will give you better control over the cooking process as the same rule applies—LESS HEAT TO REACH THE SAME DONENESS.

Buffalo meat, cooked properly, tastes very much like beef, only better! Buffalo has a fuller, richer flavor. It is not gamey or wild tasting.

If you have reservations because you are afraid buffalo are endangered and if you eat one you will be contributing to the animals' demise, put those fears to rest. Buffalo have not been endangered for many, many years. Considering the 75,000+ (and there will be more by the time you read this) buffalo in the United States and the 16,000+ in Canada, you are not committing an ecological sin when you eat some. What you eat is doubtless from a surplus bull culled to prevent overgrazing the range—so you are doing your ecological duty when you eat him.

Buffalo meat is available today because there is a market for the product. Every time you purchase buffalo, the market grows. This encourages the production of more buffalo and gives ranchers and farmers the incentive to convert their operation to raising buffalo.

The dramatic increase in the number of buffalo in this country in the past 30 years is directly related to private enterprise and profit. As more people demand buffalo meat, more people will raise more buffalo. It's that simple.

For data on the properties of buffalo meat and taste-tempting recipes, order the NBA's *Buffalo Cook Book* for $1.75, plus $1 postage and handling.

HOW DO YOU COOK A BUFFALO?

by Pat McNees
Washington Post

I can walk past a high-ticket item for years and never be tempted, but once it's sold out I tend to want it urgently. Thus with buffalo. Never mind that a student in my daughter's class gave a book report on buffalo as an endangered species (which it no longer is). Never mind that my husband said he couldn't imagine eating something about which he had sung "home on the range" so many years—a national mascot with a mystique of majesty. Never mind my friend, Julie, who said that she would blackball buffalo for dinner as belonging somewhere between beef and Bambi.

At Safeway recently, I had a brief glimpse of buffalo in the meat section, at a price slightly higher than beef steak. By the time curiosity overcame frugality and brought me back to buy it on another day, I learned that it was all sold out. Safeway's local meat buyer, Henry Hughes, explains that they buy buffalo about once a year, whenever it's offered to them, have it slaughtered and put in selected stores. The 150 or so head they bought this year (1981) didn't last long, even at $4.99 a pound for porterhouse and T-bone steaks, $3.79 for sirloin, and $1.99 a pound for ground meat.

I located, and contacted, the National Buffalo Association, a confederation of buffalo buffs and growers who think the only thing more dangerous to the current renaissance in buffalo ranching than government regulation is the handful of bleeding hearts who think killing a buffalo is a rape of the planet on the order of killing whales and baby seals. "The main reason the buffalo has come back from near extinction is because private enterprise found a way to make a profit on the animal," writes Judi Hebbring, executive director of the NBA and editor of *BUFFALO!*, one of the liveliest specialty magazines in circulation.

Buffalo were once at home on the range. Bison bison, as opposed to, say, the water buffalo of Asia or the cape buffalo of Africa, are more commonly known as just bison. They thrived on the protein-rich American grasslands during westward expansion. But as the market for buffalo hides and tongues grew, the ranks of Bison bison were decimated. A few private individuals recognized the danger to this noble animal and protected the remaining scattered herds until greater public awareness led to the creation of national and state parks to preserve the species.

By the 1920's and 1930's, the public herds began to grow. Today, there are about 75,000 head of bison—a big jump from the turn of the century, but insignificant when you consider that we slaughter more than that many head of beef cattle in a single day in the United States. Only 5,000 to 9,000 buffalo go to slaughter annually. Most of the animals are used as breeding stock, and herds are building up all over the country—some of them within commuting distance of downtown Washington, DC.

With 68 head of buffalo, Wray and Roma Dawson of Chantilly, VA, have one of the largest herds in this neck of the woods—small potatoes compared to some of the large herds in the West, but symptomatic of a trend the National Buffalo Association encourages. The Dawsons started seven years ago with 10 head: nine cows and a bull. Why did they switch from raising cattle to raising buffalo? "We've been in the cattle business for years," he explains, "and we've lost money every year. You can get maybe $400, $450 for a Black Angus, and the rancher has nothing to say about that price—he's strictly at the mercy of auction prices that day. A bison on the hoof will bring you $1,000—and you aren't controlled by the auction."

The Dawsons also have no trouble selling what they slaughter, despite prices one would hardly call giveaway; and in addition they sell the unprocessed hides for $75, the hooves for $10 each (to Western art galleries), and the skulls—skinned and boiled—for $75, to Indians, who use them in the ceremonies.

"Not only does the Department of Agriculture have no information about buffalo," says Wray Dawson. "In fact, they have a bias—they think it's a wild animal. Which is true."

Of the five types of buffalo that exist in the world, only Bison bison has resisted domestication, and that seems to be part of its charm for people like Wray Dawson. "We've bred the survival instinct right out of cattle," he insists, "but not out of the bison . . . You don't pet a buffalo." You don't milk them either—not the ones in America.

Mainly you get to know what makes them tick and act accordingly. You don't run buffalo on two sides of a fence, for example, or you soon have one less fence, so strong is their herd instinct. They are so gregarious, in fact, that if you don't have enough animals to form the semblance of a herd you'll have all you can do to get them to stay. You keep away from bulls in rutting season and mothers with calves ("Buffalo are supermother," explain the Dawsons), and if you see that a buffalo is mad, you give it plenty of room.

One of the buffalo's chief advantages over beef cattle is dietary: Because buffalo convert their food efficiently, you can produce more buffalo meat at less expense than it takes to produce the same amount of beef—and get higher prices for it too. Mainly what this means is that you can produce marketable buffalo meat without the expense of the grains routinely fed to beef to increase the marbling American customers have come to expect. What's more, buffalo live more than twice as long as cattle and continue bearing young a long time; their average life span is 25 years, but many live to be 40 and continue bearing young until they die. Attempts to crossbreed cattle with buffalo—to produce the mildly bally-hooed "beefalo"—have failed because the hybrid breed won't hold true.

American consumers have not responded enthusiastically to the texture of grass fed beef, but grass fed buffalo is both leaner than grain fed beef and equally tender if cooked properly. It's nutritious, high in protein, and relatively low in fat—not so low as the buffalo industry might suggest, perhaps, but as red meats go, more than respectable. According to researcher Wayne Johnson at South Dakota State University, who has just completed a study of the nutritional content of buffalo, the edible portion of grass fed buffalo is 4 to 5% fat, whereas similar cuts of grass fed beef contain 6 to 7% fat. It is not a huge difference, but as buffalo ranchers are quick to point out, the most useful comparison would be buffalo versus the higher fat, grain fed beef. Any claims you hear that buffalo are cholesterol-free are untrue, but because of its low fat content, grass fed buffalo do have only 150 to 200 milligrams of cholesterol per pound of cooked lean meat—enough to make it safe in moderation in most low-cholesterol diets.

The hidden phrase in the above discussion is "if cooked properly." Cook buffalo at too high a heat or for too long and you'll get the taste and texture of shoe leather. Cook it at low heat and not too long and you'll get a tender, juicy piece of meat with a taste not unlike that of beef—and not at all gamy, by the way.

Johnson's experimental team ran kitchen tests of various cooking methods and found that consumer tasting panels gave the highest ratings of buffalo roasts cooked in an open pan, in dry heat, at 275 or 300 degrees to an internal temperature of 155 to 165 degrees (rare or medium-rare). We followed the Dawsons' instructions on the meat we tested—prepared with utmost simplicity, so we knew we were tasting essence of buffalo—and got a very positive response from dinner guests who arrived curious but slightly skeptical, and departed fans. Here are the Dawsons' instructions:

HOW TO COOK BUFFALO

The first time you try buffalo, keep the cooking simple. It is very important to realize that buffalo muscle fiber is finer than beef, and therefore cooks more quickly. So cook it at a lower temperature than beef and for a shorter time. Buffalo meat is richer than beef, so it can go further.

ROAST BUFFALO

Sprinkle your roast with lemon pepper (don't overdo it), then place a couple of strips of bacon over the roast and cook in an open pan at 325°, 15 to 20 minutes per pound. Melt a little butter and pour over the roast about midway through roasting.

STEAK

Fix the steak as you would beef. Just remember to cook a shorter time.

GROUND MEAT

Use your favorite recipes. If using as burgers, just remember to cook only until the pink is gone.

25
Amazing Healing Properties of the Buffalo

Buffalo possessed for the Plains Indians mystic healing powers. Fresh buffalo dung poultices, for example, were used for ragged, infected wounds. An unlikely remedy, yet according to latter-day medical evidence, it worked.

You can't find a rancher who ever saw cancer in buffalo—and ranchers are all too familiar with cattle cancer. Nor have we been able to find reference to buffalo cancer in scientific literature.

Through the buffalo's great arteries pulse healing miracles. Mankind is seeking—but only barely suspects—how far reaching they may be. It is possible this majestic animal offers cures for countless people all across this planet!

DO BUFFALO PROMISE THE CONQUEST OF CANCER?
by Dr. Dwain W. Cummings

Note: The following is based on conversations and correspondence with Dr. Cummings and upon information in the books his wife, Pinky, wrote about him and his work:

Why They Call Him the Buffalo Doctor

by Jean Cummings. Prentice-Hall 1971

Alias the Buffalo Doctor

by Jean Cummings. Swallow Press 1980

(Dr. Cummings has been an active NBA member continuously since 1967.)

"Buffalo don't get cancer." You've doubtless heard buffalo raisers say this. Sounds like an Instant Advantage to any experienced raiser of cattle or horses, for cancer losses are a real burden. Picked up at an NBA meeting by a new rancher, this bit of folk wisdom fired his interest in the immunologic approach to cancer prevention and treatment in human beings. He is Dwain W. Cummings, D.O., Muskegon, MI.

In the early '60's, associated with a small country hospital, he installed a small herd of the beasts in the pasture next to the hospital to give his patients something to think about besides their aches and pains. He named it the R_x Ranch and used the R_x brand. People started calling him The Buffalo Doctor.

Immunity

An osteopathic physician and surgeon, Dr. Cummings is trained and skilled in marshalling and augmenting the body's natural defenses to disease. If, he reasoned, buffalo are indeed immune to cancer, would a transfusion of buffalo blood serum transfer this natural immunity to human patients and help them whip cancer? Or keep them from getting cancer? The notion of a cancer vaccine flitted through his busy brain. Might there be some way to enhance the buffalo's natural imm nity for transfer to the human being? But how? The human body usually reacts violently to any foreign protein injection.

Then he picked up another tantalizing comment from NBA members: "Nobody's allergic to buffalo meat."

Allergies, Dr. Cummings explains, are immunities gone wild. People become immune to pollens, dusts, milk, certain kinds of meats, etc., and the body tries to reject them.

The Buffalo Doctor further explains that the body manufactures antibodies in response to disease germs. Antibodies are blood proteins that destroy invading germs. The body may already have antibodies against certain diseases, but when attacked by something else—the measles virus, for example—it gets the disease once, throws it off, and meanwhile manufactures the antibodies that keep it from getting the disease again. Now that person is immune. Inoculating (vaccinating) with killed or weakened disease germs stimulates the body to manufacture antibodies against that disease: "active" immunization—long-lasting, sometimes permanent.

Dr. Cummings also described "passive immunity" for us: " 'Passive immunization' is conferred by injecting the patient with blood serum from an immune individual. Passive immunity helps the patient throw off the disease, or at least it makes the sickness milder and speeds recovery. Examples are diphtheria, polio, and poison snakebite. Passive immunity is usually temporary, although in some cases it 'programs' the body's immune system to manufacture that, or a related, antibody."

Non-Allergenic, Too!

The Buffalo Doctor investigated the allergenicity of buffalo meat, milk and blood serum. "All proteins," he said, "including buffalo proteins, will

320

produce allergic reactions if an individual has been sensitized. This is part of the basic characteristic of proteins.

"Buffalo proteins will produce allergic reactions *only* if the person has been sensitized to buffalo proteins. However, very few people—I have found none—have been so sensitized. Allergists prescribe buffalo meat to patients allergic to domestic beef.

"This situation is far different from horse serum," he notes, "which is much used as a lockjaw preventive and to prevent organ transplant rejections. Many people are allergic to horse serum. Horse serum can kill a sensitized patient.

"Organ transplant recipients may have to receive horse serum injections for the rest of their lives to keep their bodies from rejecting the new organ. But some of these people become violently allergic to horse serum. Too, horses are highly susceptible to many forms of cancer, and some of these transplant patients develop a cancer at the horse serum injection site.

"On the other hand, we can find no instance of a buffalo in the wild state having succumbed to cancer. We must develop a non-allergeic serum. All the above information points to buffalo blood as being a good non-allergenic base for such a serum."

These facts all fitted together, making Dr. Cummings' idea look all the more practical. They added fuel to his research fervor.

More Cowboy Than Scientist

But buffalo are not white mice. They are huge, wild, powerful, fierce, dangerous. How do you give a herd of buffalo their fortnightly injections and then take repeated quarts of blood from them?

The Buffalo Doctor therefore chose a buffalo calf to begin his experiments, since a calf is easier and safer to handle. "Crazy Horse" was his name . . . Hoss for short. Doc injected increasing doses of human white cells into Hoss at two-week intervals. Pinky helped him. It was like a fortnightly rodeo in their suburban back yard (The Little R_x Ranch), "Our part in this research calls for more rodeo skill than scientific prowess," Jean grunted to her husband as Hoss kicked and gouged her ribs with his little spike horn.

"Be careful!" warned her husband solicitously. "Don't break that horn!"

After the fourth injection, Hoss staggered out of the chute on his knees. "An anaplasmatic reaction!" Doc exulted. "Proves Hoss is building antibodies against the injected cells!" But Hoss' temperature shot to 106° (buffalo normal is 100°). Doc feared he would die before dawn and sat up with him all night, but Hoss snapped out of it.

Hoss' reaction proved the first part of Doc's theory. Now to the second part: Could these antibodies be transferred to human transplant patients without all the serious side effects of horse serum?

Crazy Horse became increasingly difficult to handle as the experiment progressed.

Doc and Pinky drew blood samples from Hoss at one-week intervals. Their hospital laboratory centrifuged off the serum which they stored in their kitchen deep-freeze. When they accumulated four liters, they air freighted the frozen fluid to a research institution.

Pinky relates, "The laboratory technicians were amazed at the sky-rocketing antibody titre level Crazy Horse was showing." Electrophoresis indicated a titre twice as high as that obtainable in the horse, and in half the time. "Buffalo did indeed have a unique and violent immunity response mechanism!" she notes.

But the blood withdrawals were too much for Hoss. He stopped growing, became wrinkled and old looking. Hoss died of old age before he was a year old.

A Friendly Ear

Pinky explains that Doc believes we humans have a strong natural immunity to cancer as compared to our immunity to other killing diseases. Untreated bacterial infections can kill you within hours, but cancer needs months or years to kill. This has to be because the human immunity mechanism, Dr. Cummings believes, is closely related and responsive to that person's spirit. (Many physicians have noted that some of their cancer patients appear to have given up on life even before this dread

322

disease hits them.) This immune response is similar, Pinky notes, to the fight-or-flight response when our glands inject adrenalin into our bloodstream in danger or fear. The immunity response takes hours or days to peak, instead of the split-second adrenalin reactions.

"One of the problems of working with cancer cells," Doc told me, "is that they are a poor antigen. That is, they don't stimulate the body to generate strong antibodies." Despite this, he proved it possible to inject buffalo with cancer cells from a human patient, causing the animal to develop strong antibodies against those cancer cells in eight weeks.

"By drawing some of the buffalo's blood and separating off the serum, one can isolate the antibodies and they will destroy cancer cells in another individual. The process has yet to be clinically evaluated with human patients."

He points out that this is not a true vaccination. "It is a process of instilling passive immunity temporarily. However, that individual's spirit and immune mechanism then copy the pattern for production of the individual's own antibodies. His antibodies then contain the information, in fact, to make that individual immune to that type of cancer. This information can then be used by his immune system as a beginning pattern for development of immunity to other types of cancer.

Beth and Brenda Cummings and calves selected for the 1975 experiment.

"This is a subtle, but absolutely crucial, difference!"

Working with his own funds and on his own time limited the busy surgeon's research progress. He found a willing ear in Parker B. Francis III, a director of the American Cancer Society. Dr. Cummings described to Francis the uncommon immunity mechanism of buffalo.

"Buffalo are unique," he explained, "in that they have either a complete immunity to cancer, or at least a very high degree of immunity. In the years I've been chairman of the NBA Research Committee, I've seen only two cases of cancer in buffalo. Both of those were confined, penned up in zoos." Francis worked with the doctor to establish a funded research project, but was killed in an auto crash before anything came of it.

"The lab people were very enthusiastic to go on and sew it up," Dr. Cummings reports. "The agreement was they were to send the lab data to me so I could make a solid convincing report or publish the results. But they 'lost' the data. When I went back to the places that Mr. Francis had interested, they were interested only in the reports which I didn't have."

That's how close humanity came to conquering cancer.

Jenner, Salk & Cummings

Dr. Cummings' discovery, when put to clinical use, will rank with Jenner's smallpox vaccine and Salk's polio preventive. Yet The Buffalo Doctor is not the least concerned that history record Dr. Dwain W. Cummings as saviour of the human race. What he wants is the conquest of cancer—for someone to carry on the work he began and has not the funds, nor the facilities, to finish.

Vital New Job For Buffalo Raisers

When the disease-fighting power of buffalo serum is clinically recognized, buffalo people will find themselves fullbacks on the world health care team along with researchers and physicians. For it will not be laboratory technicians in neat white coats that manage the buffalo herds to produce anti-cancer serum. Ranchers in patched pants and muddy boots will inject them and draw the blood.

Because buffalo cannot abide confinement, they will either "will themselves to death," as Buffalo Jones reported a century ago—or succumb, as Dr. Cummings observes, to a broken spirit. And it is this wild spirit of the buffalo, The Buffalo Doctor believes, that makes buffalo potentially so valuable to humanity. "Lock him up and break his wild spirit," Doc warns, "and the buffalo becomes no better to us than a cow."

The Buffalo Doctor has shown the way to cure cancer—prevent cancer—control organ transplant rejection—have a more nutritious diet. He's made all these breakthroughs.

Is there no one to carry on?

DOCTOR THOUGHT BUFFALO TALLOW MIRACULOUS

Following is some correspondence from the early 1900's between Col. Charles Goodnight—one of the ranchers who helped save the buffalo—and a Los Angeles physician.

Office of
C. GOODNIGHT
Breeder of
BUFFALO, POLLED ANGUS CATTLE, CATTALO AND KARAKULE SHEEP

Goodnight Texas, Aug. 31, 1916

Mr. Edmund Seymour, Pres.
    American Bison Society
        45 Wall Street, N.Y.

My dear Sir:-
        Yours of the 28th reached me this morning, and I answer in haste, as I fear I did not make myself clear in the advice given you in regard to buffalo soap. It is not soap at all, but so made. It is made into soap with lye. Now, we take this soap and mix it with water and it makes a compound with far reaching qualities. It has no taste or smell of soap, and has qualities unknown to us. Have patience and I will enumerate things it does.

        I put two applications on a corn and it absolutely stopped soreness. I had a lame knee and it cured it. One of the tenants was lame in the back and a few applications cured him. In applying it to the skin it leaves no grease. It will clean oil paints and not injure, clean silver and books of all description. You can take the most delicate fabric and wash in it, and it will not fade. It is a perfect disinfectant. If you have grease on your hat it will take it out. It leaves flesh soft and without injury. It seems to kill all insects. We had a cup containing a number of red ants; we took our finger and rubbed it in compound and rubbed in around top of cup. In a little while every ant was dead. It could not be of any great value as soap. Get the dirtiest old painting you can find, spread this compound very thinly on picture and frame, let it stay five minutes and rub off with soft cloth. I am satisfied it will relieve rheumatism. By all means have it tried on the infant paralysis. Try it for tuberculosis. I do believe it will work. It is harmless. We do not know what we have found. Help me hunt it out. I believe it stands a fair chance to become the discovery of the age.

        We send you several pounds by express, prepaid, today. Please do not treat this as bosh. Give it a test in behalf of humanity. If you will do this you will find its use unlimited. The Pueblo Indians have believed for centuries that it is an antidote for tuberculosis. Give it a trial. We would like to know as much as the Indian.

      I remain,
           Yours very truly,

           C. Goodnight

Goodnight, Texas, Sept. 30, 1916.

Mr. Seymour,
My dear Sir:-

Your letter to my husband is just at hand, and as I know so much more about the Wild Buffalo Balm than any one else, I write this myself. You tell your doctors from me that I have been working over the balm for six or eight weeks and know whereof I speak, and came about in this way. I had more buffalo fat than common; and made it into hard soap by cold process in this way by directions of Babbitt's concentrated lye - six pounds of clean grease, one quart of water with lye dissolved in it, called one lb. It takes about thirty minutes to get it thoroughly mixed and perfectly smooth, about like honey. Let it set for ten or twelve hours to suit your convenience. I generally let it set till next day. I had made this kind of hard soap all my life with any kind of clean grease. This year I notice several differences in this and other hard soap. One was it would form a jelly-like substance if you made a strong suds and left it a few hours. Another thing, it would not fade anything washed in it, hot or cold water. Then I noticed its cleaning properties - beat anything I ever saw, so I commenced cleaning some old plated silverware almost black with age and neglect, until I use it every day on my table and looks like new. I dipped a brush in the compound and rubbed thoroughly, and the dirt went off like magic, and after rubbing with a dry cloth clean and dry all the dirt disappeared, and the most wonderful polish came on, so I continued make the compound and stirring out the soap and boiling well till thoroughly dissolved. I kept it on the fire boiling hot and put boiling clean water in it until it is still a mystery to me how much it would make of a kind of very white gelatine. I weighed one pound of the soap and made three gallons of this gelatine. I know I could make five gallons thick enough for varnish or shoe polish. I put it on every photograph and cleaned them perfectly, polished better than they had ever had. I cleaned my old oil paintings with it perfectly, frames and all, and I had never dared to put soap of any kind on them before. After adding so much water, the stuff has no taste or smell, only a flavor of soap is all I can see in it. Tell the doctor that the grease I have kept sitting in an open vessel does not spoil in nearly two years. I have had some of it and flies don't get in it and they do in everything else about my place, for they swarm about the horse loft near the house. I have worked so constantly with the stuff I cannot think of much of anything but the possibility of doing something great with the buffaloes for humanity.

Sincerely yours,

Mary A. Goodnight

Office Hours
9 A.M. to 6 P.M.
Sundays 9 to 12

H. J. TILLOTSON, M.D.
304-305 O. T. Johnson Building
Broadway at 4th St.
Los Angeles

Apr. 22nd, 1918

Col. C. Goodnight
Goodnight Texas

Dear Sir and Friend:

Since you forwarded the tallow to me I have used it as I wrote you I would. However you no doubt know that I am a Specialist and do not cover the entire field of medicine. On this account I have requested of my friends in the profession who practice other lines that they try this preparation in their practice and I am to get their reports from Clinical cases.

I treat many skin diseases along with other work of my regular line which includes diseases of the alimentary canal and the pelvis. I have incorporated the tallow in suppositories for use in Rectal cases - in addition to this I have used it straight in severe cases of Pruritus. I report great benefits in the cases upon which I have tried this.

Having to supply other members of the profession as I have herein stated I have none of the tallow left - it is my intention to go the limit in this trial and will therefore ask that you send me a large quantity - say twenty pounds. I will be pleased to pay you for the tallow if it is not too expensive. This amount will permit me to use it more freely which is the only way this problem can be solved.

Trusting I may hear from you in the near future and with my best wishes, I am,

Sincerely yours,

H. J. Tillotson, M.D.

A Glimpse into History

VII
VII
VII
VII
VII
VII
VII
VII
VII
VII
VII
VII
VII

26
The
Noble Beast
Called Buffalo

To try to tell the saga of the American buffalo in a few pages, one must write fast. Scientists specializing in such stuff think *B. b. bison* descended from the European wisent and migrated along with the early Indians across the land bridge that is thought to have connected Siberia and Alaska some 50,000 to 800,000 years ago, depending on which authority you read. (The wisent, too, came within a breath of extinction during food short WWI, and again in WWII, but was saved like the American bison, more by concerned private citizens than by government.)

The American ancestor of the buffalo was a huge beast with nine-foot horns. It is thought the Indians hunted him to extinction, but not before he gave rise to the present species. Pioneer accounts of buffalo seem incredible to us: herds extending from horizon to horizon, taking days to thunder past an endangered campsite, derailing trains and stopping Missouri River steamboats by their sheer numbers. Estimates of the total buffalo population as discovered by Europeans range from 60 million to 120 million (about the same number as cattle in the U.S. today).

The Plains Indians suddenly graduated from a dog culture to a horse culture, mastering the horse in less time than it took us to master the gas engine. By the time of the Revolution, Plains Indians were roaming wild and free on horseback, supplying most of their needs from the buffalo herds which they followed with prayer, charms, and musket.

Not only did the flesh give them food and the hides give them clothing, shields and tipis, but just about every scrap of the carcass was used for practicality, and the buffalo spirit for religion (see "Buffalo" chart).

After the Civil War, courageous hunters discovered it was great fun to slaughter the beasts by the hundreds, leaving them to rot. The hide hunters killed more millions for the $1 hides, and the railroads ran special trains for the pleasure and edification of sportsmen who shot from windows of moving trains.

The couple of decades of the commercial hide hunt has been well-documented ... meat rotting on the plains while jobless Eastern families starved. Many Indian troubles were sparked by the wanton destruction of their livelihood. Indians found it difficult to understand why an Indian should be hanged for shooting a settler's cow to feed his starving family while the settler could shoot hundreds of the Indian's "cattle" for fun. Buffalo extermination became official government policy to destroy the Indians' base of supply, thus starving them into submission.

After the hide hunters got through, it was reported you could walk for miles on the carcasses without ever touching the ground. The stench and the flies made eating difficult.

Suddenly, in 1883, the elaborately-equipped hide hunting expeditions returned with empty wagons. Inconceivably, the inexhaustible buffalo were gone! When the National Museum realized they did not even have a single decent mounted specimen, they sent out an expedition to get some. The 1886 foray almost didn't find any.

Following the hide hunters, the wolfer had a short day, cashing in on the plenteous pelts of these prairie scavengers. Then came the bone pickers. Homesteaders had to pick up the bones before they could plow. The bones, delivered by the wagonload to the railroads, brought small sums that kept many a homesteading family alive during that difficult first season while waiting for a crop. Newspapers tell of bonepiles many feet high, many feet wide and half a mile long beside the tracks, awaiting shipment to the fertilizer factories of the East.

The low point of the buffalo population was reached in 1900. Nobody knows how many were left—but it's pretty certain the world population was under 1,000 that year: A few in Yellowstone Park, falling to poachers' guns; a few in zoos and private estates here and there; a few small herds on ranches; some wood buffalo in Canada.

Salvation apparently centered in several spots. Walking Coyote, a Pend d'Oreille Indian in the Mission Valley of Montana, saved four calves, two bulls and two heifers—and drove them home in 1873. These became the foundation of the famed Pablo-Allard herd in Montana, which in turn restocked Canada. C. J. Buffalo Jones, Garden City, KS, captured a few to try crossbreeding and domestication experiments. Charles Goodnight, down in the Palo Duro Canyon of the Texas panhandle, saved some. The Indian wife of Scotty Philip in central South Dakota begged him to save some, so in 1901 he bought the remnants of the Fred and Pete Dupree herd at Dupree, SD. Scotty's herd eventually provided foundation stock for the Sutton herd of Onida, SD, and the Custer State Park herd. From these tiny centers sprang all the 75,000+ head now begetting their kind throughout America, and more thousands in Europe.

Had it not been for pioneers like Walking Coyote, Jones, Philip, Goodnight and a few more, probably the buffalo would have gone the

Buffalo
(American Bison)

The parts of the buffalo not consumed as food were put to some other purpose: clothing, tools, weapons, furnishings, and ceremonial objects.

The bison were the Indians' All-purpose Supply Store.

LifeSpan: 30 years

Ears: Keen
Eyes: Weak

Maturity: 6-7 years
Weight: up to 2000 lbs.
Nose: Buffalo's grass-locating and danger-warning instrument

Jumping Ability: 6 ft.

Speed: 35 miles per hour, with more stamina than a horse

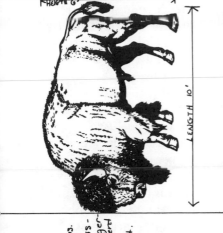

LENGTH 10'

	Nourishment	Clothing	Home Furnishings	Tools and Utensils	Weapons	Transportation	Recreation	Ceremonial Objects
EDIBLE PARTS	Meat (hump is the delicacy) Tongue, Blood, Brains, Heart Intestines (while hunters craved these) Fat (pounded with dried meat and berries to make pemmican) Stomach (including greasy contents)			Bladder (drinking vessel) Hairbrush (rough side of tongue) Soap (fat) Tanning agents (brains and liver) Patches (intestines)				
HIDES		Robes Shirts Leggings Dresses Belts Moccasins Caps Mittens	Tipis Blankets Medicine cases Trunks Cache-pit covers	Tobacco pouches Berry bags (hide of unborn calf) Cooking vessels Buckets	Splitting Knife sheaths	Horse coat Dog gear (before horses, dogs were the pack animals) Saddle bags Lariats Horse-watering troughs Boats, rafts Snowshoes	Ball covers Netting for lacrosse Hoops	Rattles Masks Winding sheet for dead
HORNS		Headdress ornament.		Spoons, ladles, cups	Powder flasks			Masks
SKULL				Rawhide rope was pulled through eyesocket to dehair it				Altars Entire head worn as mask
BONES		Ornaments (used teeth also)		Paintbrush handles Knives, scrapers Hoes (shoulder blade) Sewing awls	Arrowheads Lance points	Sled runners (rib bones)	Counters in gambling games	Masks, rattles
HOOFS				Glue	Ornaments for clubs			
HAIR				Brushes Braided into rope	Knife sheaths		Stuffing for game balls	
TAIL				Flywhisks				
SINEW				Twine thread	Bowstrings Bow backing Arrowhead wrappings			
DUNG				Fuel Placemakers				

way of the "limitless" passenger pigeon and other "inexhaustible" wildlife.

William T. Hornaday, taxidermist for the National Museum, waged a decades-long fight for the protection of all threatened wild species. He, more than any other person, is credited with saving the buffalo. Largely from his efforts, Teddy Roosevelt in 1908 persuaded Congress to establish the Moiese Wildlife Refuge in Montana. A year later it established another, the Wichita Mountain Refuge, at Cache, OK.

It is notable that governments did little—and did it too late to save the herds. The Canadian government bought 600 from the Pablo-Allard herd in 1906 after the U.S. government had repeatedly refused.

Today, with the species safely off the endangered list—thanks mostly to buffalo ranchers—the public can best encourage them to aid the spread of the species by buying buffalo meat and other products. One need have no fear of committing an ecological crime because it is only by utilizing the surplus animals that ranchers can be enabled to rebuild the numbers of this noble beast which best symbolizes America's pioneer past.

BUFFALO IN THE EAST

condensed from an article by
Duncan A. Dobie
Marietta, GA

It is difficult to imagine a Cherokee brave, bow and arrows in hand, stalking buffalo through a North Carolina hardwood forest. Many of our early pioneers hunted buffalo in Kentucky and Tennessee.

When white men arrived in the New World, the buffalo's range in the East extended from New York to northern Florida, and southward from the Great Lakes into the Ohio Valley and west to the Mississippi River down to the Gulf coast. Herds were never as large as in the West, usually ranging from a few animals to several hundred.

Buffalo were as valuable to the Eastern Indians as to those of the Plains: furnishing food, clothing, shelter, weapons, utensils, decorations, and ceremonials. Many warriors of Southeastern tribes used armor and shields of tough buffalo leather. The Choctaws of the Gulf coast made an ingenious cylindrical fish trap out of fresh buffalo hides. Blood from the hides attracted catfish, and when a fish swam into the trap, the fisherman pulled a drawstring and closed the trap. Southern Indian women dyed buffalo hair which they wove into patterns in their textiles.

Networks of buffalo trails aided settlers and Indians alike in their travels. Many of today's cities stand where buffalo trails led.

A section of the Ohio River basin was said to be the home of the legendary "Pennsylvania Bison," a coal-black variety residing in dense forests and adjoining grasslands. Many thought these animals to be a separate sub-species, but this was never proved.

Although the Eastern Indians engaged in large-scale trading of deer hides with the whites, little evidence has been found that they traded much in buffalo hides. This may be because they were much scarcer than deer hides and therefore too valuable to trade. Some were given by chiefs as prestigious gifts to white leaders.

Already by 1700, buffalo numbers were rapidly declining in the East. By 1750, few were being killed. Both Indians and settlers expressed concern over the dwindling numbers. In 1759, Georgia legislated protection for "deer, beavers and buffaloes" in certain areas. People seemed to think settlers and livestock were driving buffalo out; they did not realize the animals were being killed off.

By 1800, only a handful remained in the East; by 1820, none. In Ohio, they were wiped out by 1802. Pennsylvania's last herd was slaughtered in 1799, Wisconsin's in 1832.

The extermination of the buffalo in the East was probably not as deliberate or methodical as in the West, but resulted in random hunting by hungry pioneers. They also exterminated the eastern elk, timber wolf, black bear, cougar, wild turkey, passenger pigeon, eagle, and other raptors. Whitetail deer, squirrel, and raccoon were all brought to the edge of extinction in the East. Many of these species are once again thriving, and today we hardly realize how close some of them came to the brink.

BUFFALO WERE FROZEN FOOD FOR INDIANS

An ancient "frozen food locker" gets very careful attention each summer from a University of Wyoming archeological team working under the direction of George C. Frison, head of the UW Department of Anthropology, who is also a Wyoming State Archeologist.

Frison, a team of 18 to 20 students enrolled for a summer field school, and several specialists continue excavations at the Agate Basin Site. This is a particularly rich lode of bison bones and artifacts deposited by three distinct cultures of nomadic hunters active as early as 11,500 years ago.

Evidence uncovered and classified to date has revealed the remains of more than 100 extinct bison mixed with killing and butchering tools in a series of locations scattered over more than 300 acres. The site is near Mule Creek Junction between Newcastle and Lusk and within about a half-mile of the South Dakota border.

Frison, who has been gathering data and artifacts at the site for many years, concludes that the location was occupied more or less continuously for several years by bands of Clovis, Folsom and Agate Basin hunters. These Paleo-Indian cultures are identified by their distinctively formed stone projectile points.

"What we've found so far strongly suggests that these early hunters made mass kills in the early winter, allowed the carcasses to freeze, then removed parts over a period of time as they were needed," Frison says.

THE BUFFALO JUMP

GTA Journal
January, 1982

High above the small band of Indians, on a plateau that abruptly ended at a cliff, came a distant rumbling. The closer and more intense the sound of hoofs pounding against dry earth came, the greater the Indians' excitement. Soon a cascade of bison, like a waterfall, would plummet over the brink to their deaths, 20 or 30 feet below.

Of the many methods horseless Indians used to kill bison, perhaps none was more dramatic than these "buffalo jumps" . . . sometimes called "pishkuns."

The pishkuns, a Blackfoot Indian word meaning deep-blood-kettle, were an efficient way to gather needed supplies. In addition to meat, one researcher estimates that Indians processed 87 products for their use from bison carcasses.

Used as early as 2,000 years ago and as recently as 200 years ago, hundreds of buffalo jumps have been found, primarily in Rocky Mountain areas, from Colorado to Canada. At these jump sites, usually a ridge-rock cliff carved out of the plains by a river or stream, archeologists have found deep deposits of bones and evidence of slaughter sites used by the nomadic tribes to process their kill.

Although buffalo jumps provided tremendous amounts of meat and hides quickly—sometimes as many as 100 bison a kill—the hunt demanded patience and skill. While Indian hunters formed a V-shaped funnel leading to the cliff, a young, swift Indian with intimate knowledge of bison would sneak out and gently move a herd towards the trap by exciting lead animals.

As the herd moved closer, Indians dressed in robes fell in behind the bison, yelling and creating enough commotion to stampede animals to the precipice. At times lead animals would sense their impending fate, but penned in by hunters on both sides, there was little chance to escape. They and their comrades were pushed over the edge by their own momentum.

After any crippled survivors were killed by hunters waiting on the slopes below, the hard work began. The cliff's base became a small industrial site. Animals were dressed, tons of meat were dried, fresh meat was consumed for several days, bison skins were tanned, and other necessary parts of the bison's blessing were taken. After the meat was dried, the nomads moved on, leaving waste for scavengers. In two or three months the site would be used again.

The Shoshone, Blackfoot and Crow Indians were all said to have used buffalo jumps to fill their larder and provide clothing and shelter. Fall was probably the most popular season to use jumps, as large quantities of supplies could be readied for harsh northern winters.

The Buffalo Jump was an early killing method used.

FAMOUS BUFFALO HERDS – I

by Dave Raynolds

There were only a few buffalo herds in 1888–89, according to records compiled by Dr. William T. Hornaday, then chief taxidermist of The Smithsonian Institution. Hornaday, who later became a buffalo conservationist, feared that buffalo were dying out in the 1880's, and wanted to stuff a group while he still could. In his report, *Extermination of the American Bison*, published by the government in 1888, he recounts how he finally found a wild herd between the Yellowstone and Missouri rivers in Montana, and succeeded in killing 24 of the 25 animals sighted.

The following listing of herds in that period draws on his work and a later book by Martin S. Garretson, *The American Bison*, published in 1938 by the New York Zoological Society. Garretson was Secretary of the American Bison Society at that time.

For January 1, 1889, Hornaday estimated 1,091 animals. This included an estimated 550 in Athabaska, Northwest Territory, Canada. Of 534 in the United States, he estimated 200 were in Yellowstone Park. Of the remaining 334, he estimated 85 were in wild herds as follows:

In the Panhandle of Texas 25
In Colorado 20
In southern Wyoming 26
In the Musselshell country, Montana 10
In western Dakota territory 4

Like today, almost all the herds recorded in those early days were in private hands. By contrast, in 1929, of the recorded 3,385 animals, 2,489 were owned by government units.

Hornaday recorded the remaining animals as follows (pp. 458–464 in his *Extermination*):

C. J. "Buffalo" Jones, Garden City, KS: 24 males, 33 females, total 57. Jones (a member of the Buffalo Hall of Fame) had recently purchased an additional 83 head from Mr. S. L. Bedson of Stony Mountain, Manitoba, Canada, a group developed there from an original bull and four heifers purchased in 1877. Counting this purchase, Jones' total in 1889 was 140; this includes eight from Bedson he sold in 1888 to Austin Corbin for the Corbin Park in New Hampshire.

Charles Allard, ranching on the Flathead Indian Reservation in Montana, had 35 head. He got his original stock of 10 from Walking Coyote in 1884, in partnership with Michael Pablo. When Allard died in 1896, the herd of about 300 was split in half. The Pablo animals were later bought by the Canadian government to mix with their Wood Buffalo; 709 were shipped north over three years. Animals from the Allard portion went into the Moiese Buffalo Preserve, Yellowstone Park, and other locations.

W. F. "Buffalo Bill" Cody kept bison as part of his traveling circus. In the winter of 1888–89 there were 18 of them wintering on the farm of General Beale outside Washington, DC. Fourteen of these had been purchased from Mr. H. T. Groome of Wichita, KS; they had just returned from London with four calves born there. Earlier, Buffalo Bill lost an entire herd of 20 in the winter of 1886–87 when they caught pneumonia in Madison Square Garden, NY.

Col. Charles Goodnight of Clarendon, TX, reported 13 animals at the time of the Hornaday count. He started collecting in 1878.

The Atchison, Topeka and Santa Fe Railroad kept a herd of 10 at Bismark Grove, KS, as a tourist attraction.

Fredrick Dupree (another member of the Buffalo Hall of Fame) had nine purebred buffalo and some crossbreds at the Cheyenne Indian Reservation near Ft. Bennett, Dakota Territory. He had started by catching five wild calves with his sons in 1882. This herd later passed to Scotty

Philip, another Buffalo Hall of Fame member. In turn, the Philip herd stocked Custer State Park in South Dakota and later formed the nucleus of the herds formed by the Marquiss family in Wyoming and the Houck family in South Dakota. (Each of these families has a member in the Buffalo Hall of Fame).

Other private holdings in Hornaday's record were:

Dr. V. T. McGillicudy, Rapid City, Dakota Territory 4
Mr. John H. Starin, Glen Island, NY 4
Mr. B. C. Winston, Hamline, MN 2
Mr. I. P. Butler, Colorado, TX 1
Mr. Jessee Huston, Miles City, MT........................1
Mr. L. F. Gardner, Bellwood, OR.........................1
Riverside Ranch Company, Mandan, Dakota Territory 2
Dakota (in the hands of parties unknown).....................4
Mr. James R. Hitch, Optima, Indian Territory, OK 2
Mr. Joseph A. Hudson, Estell, NE 1

Finally, Hornaday recorded these zoo animals:

Philadelphia	10
Washington, DC	2
Chicago	7
Cincinnati	4
New York (Central Park)	4
Manchester, England	2
London, England	1
Liverpool, England	1
Dresden, Germany	2
Calcutta, India	1
	34

Forty years later, as noted in "Famous Buffalo Herds—II," there were some 896 animals in private hands. The wild herd of 26 in the Red Desert of Wyoming had been harvested, as had the others Hornaday recorded as "wild." But private leaders, and government officials who listened to them, had succeeded in postponing the extermination of the American bison.

FAMOUS BUFFALO HERDS-II

by Dave Raynolds

When Martin S. Garretson, Secretary of the American Bison Society, made his 18th Census of living American Bison as of January 1, 1929, he recorded 3,385 in the United States and its territories. More than three-quarters were publicly owned by federal (1,887), state (274) or local authorities (328), with the total 2,489 located in 32 states, the territory of Alaska, and the District of Columbia.

In contrast to the present situation, where most buffalo in the United States are privately owned, Garretson counted only 896 in private hands, located in 23 different states. A listing of these private herds should be of interest to NBA members researching buffalo history in their immediate areas. The data in Table I records the material printed by the ABS on pages 25–28 of its *Report* for the years 1927–1930.

No buffalo census is ever totally accurate. NBA members who attended the spring convention in 1981 on Catalina Island, CA, will note that the herd there is not listed. Originally, there were 14 in 1924, and 19 by 1934. These buffalo had been imported to make a Zane Grey movie; later they belonged to William Wrigley, Jr. At least three of the private herds listed (the Trinchera Ranch herd in Ft. Garland, CO; the Marquiss herd in Gillette, WY and the Sutton herd in Agar, SD) still continue in the same family hands. The Blue Mountain Forest Association or "Corbin Park" outside Newport, NH continues, but the buffalo are gone.

Readers will note no listing for the pioneering Scotty Philip at Ft. Pierre, SD. He had died more than fifteen years before this census, but had his sons sold all the remaining animals by 1929? We know some had already gone, as Martin Collins purchased his herd from Rod and Stanley Philip in 1925. The Collins herd is still managed by its original owner and also was not listed in the official 1929 census. Other buffalo from the Philip herd went to Custer State Park in South Dakota and to establish the Marquiss herd in Wyoming.

Garden City, KS, founded by Col. Charles J. "Buffalo" Jones, was down to a municipal herd in 1929. Jones had died a few years before the 1929 census. Col. Charles Goodnight died in December, 1929, at the age of 93. Perhaps his age or death accounts for his herd in Goodnight, Texas being listed as owned by J. S. Staley. Goodnight was the source for the Sherwin herd of Sterling, CO. That herd also is not listed in the official census but we know it to have been started in 1918 and it has been in the same family for 64 years, occupying the same pasture for the past 40 years.

Table 1.
Private Herds, 1929

State and Ranch	Males	Females	Calves	Total
CALIFORNIA:				
Cajlou—Big Pines Camp	1	2	0	3
Los Angeles—Ramona Village Inc.	2	4	0	6
San Luis Obispo—W. R. Hearst				30
COLORADO:				
Ft. Garland—Trinchera Ranch Co.	25	60	14	99

IDAHO:				
Ashton—Island Park Land & Cattle	1	3	0	4
Pocatello—Nixon Trust Estate	1	6	2	9
ILLINOIS:				
Granville—Mrs. Cora M. Hopkins	1	1	0	2
Kankakee—B. L. Small	1	1	0	2
INDIANA:				
Fort Wayne—Estate of J.H. Bass	1		0	1
Fremont—Sprague Bros.	1	1	0	2
IOWA:				
Keota—C. A. Singmaster	0	3	2	5
Keota—J. O. Singmaster	3	8	2	13
KENTUCKY:				
Murray—Dr. Wm. H. Mason	1	0	0	1
MASSACHUSETTS:				
Beverly—Otis Emerson Dunham	1	1	1	3
West Brookfield—Herbert Richardson	4	2	2	8
MICHIGAN:				
Oscoda—Carl E. Schmidt	1	3	1	5
MISSOURI:				
Jefferson City—Lee Jordan	1	1	1	3
MONTANA:				
Richland—Walter M. Truax	2	5	4	11
Stockett—J. H. Dennis	1	1	0	2
Twodot—Wallis Huldekoper	1	4	0	5
NEBRASKA:				
Beaver City—W. T. Collings	1	10	5	16
Cambridge—W. A. Luther	1	1	0	2
Fremont—D. V. Stephens	1	1	0	2
NEW HAMPSHIRE:				
Newport—Blue Mountain Forest Assn.	7	7	2	16
NEW JERSEY:				
Alloway—Reeves Timberman	1	1	0	2
Blackwood—Louis Weber	3	3	3	9
NEW MEXICO:				
WagonMound—A. MacArthur Co.	0	4	3	7
NORTH DAKOTA:				
Minot—C. H. Parker	11	19	4	34
OKLAHOMA:				
Marland—Miller Bros. 101 Ranch				72
Pawnee—Major G. W. Lille, "Pawnee Bill"	8	30	2	40

PENNSYLVANIA:				
Allentown—Gen. Harry C. Trexler	26	41	15	82
Erie—Kiwanis Club of Erie	1	1	0	2
Scranton—Dr. D. W. Evans	1	1	0	2
SOUTH DAKOTA:				
Agar—E. D. Sutton	11	26	2	39
Alcester—Gallagher Bros.	1	4	4	9
Briston—B. A. Adams	1	0	0	1
TEXAS:				
Abilene—Chamber of Commerce	1	2	1	4
Amarillo—C. J. Pumroy	3	17	4	24
Goodnight—J. S. Staley				208
Iowa Park—Will Burnett	12	13	1	26
McKinney—Otise Nelson	0	1	0	1
San Angelo—P. W. Howe	1	2	1	4
Waco—J. B. Earl	2	1	0	3
VIRGINIA:				
Marion—Smyth County Fair Assn.	1	1	0	2
WASHINGTON:				
Colville—T. M. Wilson	1	0	0	1
Yakima—Gibson Brothers	7	36	13	56
WYOMING:				
Caryhurst—J. M. Cary Bros., Inc.	3	7	0	10
Gillette—R. B. Marquiss	1	4	3	8

27
Buffalo
Hall of Fame

HALL OF FAME HONORS GREAT NAMES IN BUFFALO

The National Buffalo Association in 1980 founded the Buffalo Hall of Fame "to honor two persons annually, one historic and one current, who have contributed to the comeback of this majestic animal and the development of the industry as we know it today."

Nominations for the historic honoree are solicited from the membership and published in *BUFFALO!* Magazine. The membership then votes at the annual convention. The current honoree is selected by the Board of Directors. Plaques are presented to the living honorees at the convention, and to descendants (when they can be found) of the historic honoree.

1980 HISTORIC HONOREE: FRED DUPREE

The first historic nominee was Pete Dupree. But after the membership voted him in, discussion arose as to who really deserved it: Pete—his father, Fred—or pioneer South Dakota buffalo rancher Scotty Philip?

Scotty was an early contender, but Lawrence Peterson noted that Scotty got his buffalo start by buying the Dupree herd from Pete's estate, so the honor went to Pete. Some quick research then showed that Pete's dad, Frederick Dupree (also spelt Frederick Dupris in some historical sources), really established the herd in the early 1880's when Pete was but a small boy. Fred caught nine wild buffalo calves in the Yellowstone River country of Montana and brought them to his ranch near Eagle Butte in what was to become South Dakota. Fred died in 1898 and Pete soon after, leaving a herd of some 100 buffalo.

Scotty bought what was left of the herd in 1900, about 83 head, from the Pete Dupree estate. As *BUFFALO!* concluded, "There's no doubt that it was Fred who first domesticated and started the Dupree herd. Fred maintained the herd throughout his lifetime and Philip took it over from the estate." The Scotty Philip herd seeded many more herds.

Fred's granddaughter, Mary Rivers, of Eagle Butte, SD, accepted the NBA plaque in behalf of the family, including more than 185 of her own direct descendants.

1980 CURRENT HONOREE: ROY HOUCK

L. R. Roy Houck, Pierre, SD, was the first living member of the Buffalo Hall of Fame, chosen by unanimous vote of the Board and confirmed by the membership in convention. A long-time rancher northwest of Ft. Pierre, Roy bought his first buffalo in 1959. Roy, Nellie and their children expanded the herd to some 3,000 head, at times the biggest herd in the world.

Roy served South Dakotans as state senator and lieutenant governor. In 1978 he was named South Dakota Eminent Farmer by the state cooperative service. His other services to the state and to agriculture read like an agricultural directory, but he is best known as "Mr. Buffalo."

Roy has set the pace for the industry and has assisted countless others in getting a start in the business. He was one of the instigators of NBA and was its founding president.

1981 HISTORIC HONOREE: SCOTTY PHILIP

James "Scotty" Philip was chosen the historic honoree at the 1981 NBA convention. His granddaughter, Billie Anne Rehborg, and her husband, Raymond, of Ft. Pierre, SD, accepted the plaque. In turn, they presented it to the South Dakota Cowboy and Western Heritage Hall of Fame in Ft. Pierre, where Scotty is also an honoree.

Philip was born in Scotland in 1858, poached his living under the guns of the gamekeepers of the landed gentry, and—as his biographer says—"learned to hunt while being hunted." He came to Kansas at age 18, then to Dakota to prospect for Black Hills gold. Failing to find gold, he adventured as a teamster, army scout, cowboy, rancher and freighter. He married a half-Indian woman and together they built a cattle empire on the Pine Ridge Reservation. The town of Philip, SD, is named for him. In 1891 Scotty and Sally founded the "73" ranch north of Ft. Pierre. He also served in the state senate.

Concerned for the plight of both the Indian and the buffalo, he tried to help them. In 1901 he bought what was left of the Dupree herd and trailed them home. In 1906 Congress passed an act permitting him to lease 3,500 acres of government land adjoining his ranch for the exclusive purpose of pasturing buffalo. He built a monumental fence of steel mesh which still retains some of his herd's descendants today on other ranches.

Scotty Philip campaigned for the U.S. government to establish a vast buffalo preserve in then empty West River, South Dakota. Custer State Park, today protecting one of the largest herds, was a result of that otherwise unsuccessful campaign.

Philip buffalo founded many herds that thrive today, including the Sutton herd at Onida, SD, and the Custer State Park herd; these, in turn, seeded many more herds.

Roy Houck, 1980 Buffalo Hall of Fame inductee.

1981 CURRENT HONOREE: TOOTS MARQUISS

Mrs. Toots Marquiss of the Little Buffalo Ranch in Gillette, WY, manages a ranch featuring a herd started in 1922 by her husband's father, Ted Marquiss. Her husband, Quentin, carried on the work and, following his death in 1964, Toots and her children took charge. The herd is maintained at 75 head as a hobby by Zero Management: "Absolutely no handling—no feeding, no vaccinating, no branding, no roundup—in as near the natural state as possible." Surplus bulls are harvested by field slaughter to keep the numbers within the capacity of the 12,000-acre range.

The herd is also unusual in being totally inbred—no new blood—the entire herd is descended from the original two females and one male. Improvement has been effected by harvesting out the lesser bulls.

Mrs. Marquiss, NBA charter member, served on the Board for three years. With Nellie Houck, she co-authored NBA's first *Buffalo Cook Book*. She also helped host the two NBA meetings held in Gillette.

1982 HISTORIC HONOREE: BUFFALO JONES

Clyde J. Jones, better known to history and his many admirers as Buffalo Jones, was born in 1844 in Illinois. At 21, he went to Kansas, worked in a fruit nursery and sold real estate. By the 1870's he became a leading buffalo hide hunter and was one of the promoters and founders of Garden City, KS.

In 1886, as the great Southern herd dwindled, he realized the "wickedness" of having contributed to their near extermination and resolved to make amends by saving the species. He caught 10 calves and drove them back to Garden City, where he conducted serious experiments with crossing and domesticating them. In 1889 his herd numbered 150 of the then 250 in private ownership. There were about 200 wild ones left in Yellowstone. From 1902 to 1905 he served as Yellowstone Park game warden and increased the herd.

Although his crossbreeding and domesticating experiments came to naught and creditors dispersed his herd to pay his debts, Jones' animals founded many more herds and did contribute, as he intended, to the salvation of the species. Says a biographer, "Today, most American herds have at least some descent from the animals bred by Buffalo Jones."

1982 CURRENT HONOREE: LAWRENCE PETERSON

Lawrence J. Peterson, Newport, NE, was chosen for his contributions to the "here and now" buffalo industry. Peterson has raised buffalo since 1968 and has been involved with them since he was a teenager, when his father first purchased a herd in the 1930's.

As the third NBA president, Peterson helped to stabilize the industry. He established the Association's headquarters in Custer, SD—oversaw the operation of that new office—pointed the association in new direc-

Toots Marquiss, 1981 Buffalo Hall of Fame inductee.

tions—and put *BUFFALO!* Magazine on a regular schedule. After his two presidential terms, he served on the Board and initiated bylaw changes that improved the working structure of the Association.

Lawrence Peterson, 1982 Buffalo Hall of Fame inductee.

28
A History
of the National
Buffalo Association

Although the beginnings of the National Buffalo Association go back only to 1966, already they are lost in the dim mists of time. No records exist of the exploratory and organizational meetings, and memories are so soon fuzzy. "It was Pat who . . . no, I guess 'twas Mike . . . in 1966 or '67 . . . well, I KNOW it was Denver, or mebbe Dallas. . . ."

The following is a groping reconstruction from old-timers' memories, sparse records and the researching of Judi Hebbring and Robin Wheelwright. NBA archives do preserve the minutes of the First Annual Meeting (1967) and onward.

The idea of a buffalo breeders' association was in the minds of many of our early members for a number of years before the seed was actually sown. As close as I can determine, the seed sprouted Feb. 21, 1966, at the Custer State Park annual buffalo sale. Buffalo owners and breeders at the sale got together and decided to form an organization attuned to their common interest. Provisional officers at the meeting were Bud Basolo, presiding; L. R. "Roy" Houck acting as vice-president and Les Price as secretary. The 70 members paid $662.50 dues that year, and the organization managed to spend $168.84.

Les, Bud and Roy met again in Rapid City on March 3, 1966, calling themselves the Buffalo Breeders Association. Roy roughed out some bylaws that summer under the name of the American Buffalo Association. Sometime before the spring of 1967 came the name National Buffalo Association.

NBA archives preserve the minutes of the first formal annual meeting (and all subsequent meetings), at Denver's Brown Palace Hotel, March 11, 1967. Twenty-nine members attended. They elected Roy Houck of Pierre, SD as president and vice-president was Armando J. Flocchini of Mountain View, CA. The NBA name was officially adopted at that meeting. Purposes of the Association were formalized as follows:

349

1) To promote buffalo and buffalo products, namely meat and hide.

2) To seek fair and equitable regulations in the control of disease and in the movement of buffalo and buffalo products, both intrastate and interstate.

3) To promote and develop recreation where it is connected with buffalo.

NBA headquarters followed the president the first few years: from Roy Houck's ranch house basement—to a Pierre hotel—to Custer State Park—to President Peterson's Nebraska ranch. It was President Lawrence Peterson who, in 1977, established NBA's permanent headquarters at Custer, SD. Judi Hebbring started as part-time secretary-treasurer. The following year she became the Association's full-time Executive Director.

In 1971, NBA commissioned Dana Jennings, then of Madison, SD, to "write a book about buffalo." The result, after four years of preparation, two years of review by the NBA Editorial Committee and two years of hang ups with the printer, was *Buffalo History and Husbandry,* which burst upon an unready world in 1978. The present volume is its sequel.

BUFFALO! Magazine was launched as a quarterly in 1972 with Jennings as editor. NBA now publishes *BUFFALO!* bimonthly, plus a monthly newsletter, *BUFFALO CHIPS,* to keep members informed. Judi has been editor since 1978.

Through its legislative influence, NBA is working to simplify the trade in buffalo and products. The organization is trying to get state and federal governments to uniformly define buffalo (in some states they're wild animals, in some states they're cattle, and in others they seem to be neither fish nor fowl nor nothing else). The NBA is also working for uniformity on inoculation and testing regulations, and to lift the blighting blanket of bureaucracy from this lively, growing industry. Legislatively, NBA is fighting for buffalo producers and has convinced government officials there is a viable buffalo industry with legitimate needs to be considered in preparing laws and regulations. NBA successfully blocked importation of cheap, inferior water buffalo meat.

Along with all this is the unremitting promotion of buffalo meat and products by educating the public to this most superior and healthful of foods. The need for education is demonstrated by the fact that many people think buffalo are extinct. Still more believe they are an endangered species and anyone who eats buffalo meat, is committing an ecological crime! They don't realize that we have to eat surplus bulls to make room for more cows and calves . . . that it is only the revenue from buffalo sales that makes it possible for free enterprising raisers to continue producing more buffalo.

In 1980 the National Buffalo Association won the prestigious "Theodore Roosevelt Conservation Award" presented annually by the Old West

NBA Executive Director, Judi Hebbring, accepting the Theodore Roosevelt Conservation Award.

Trail Foundation. NBA joined the ranks of the National Arbor Day Foundation, the United States Park Service and Theodore Roosevelt National Park.

The goals and purposes of the association as set down in the bylaws of 1967 remain unchanged. Promotion and propagation of the species, plus equitable regulation and advancement of the industry, remain the primary function of NBA's full-time office and staff.

The NBA with its permanent headquarters and sound financial base is keeping pace with the growing industry and its needs. From its meager beginning in 1967, NBA has progressed to a gross annual income of in excess of $80,000 for fiscal year 1982. Nearly $25,000 is in interest-bearing savings. These assets are utilized to provide the producer with sound management information (such as this book and *BUFFALO!* Magazine), to educate the public to the delights of buffalo meat, to assist in the promotion of the industry and the product and to sort out and unify the maze of state and federal regulations that govern all buffalo producers and consumers.

NBA prepares brochures and other educational materials which it furnishes in classroom quantities to thousands of schools all over the country, telling the story of NBA and buffalo. Additionally, the Custer office fulfills requests from everywhere for information packets.

Since 1977, when we acquired a home office, a permanent address and telephone number, we have been listed in numerous library directories. As a result, authors working on buffalo articles are able to get from NBA the accurate, updated information they need.

NBA members' dues maintain the office in its above activities, finance the publications and help support various research and fact-finding projects to benefit all buffalo breeders and consumers. Visitors are always welcome!

Work in progress at publication time includes a computerized clearinghouse of buffalo market information: Who has what to sell, who wants to buy which, etc.—and a registry plan pointing toward an eventual herdbook, such as are maintained by domestic purebred livestock associations.

Bibliography

by Jim White with assistance from the NBA office

Back in the '70's, an author tried to compile a buffalo book bibliography. He listed over 900 titles in English alone, and he admitted his list was far from complete.

Our task was to compile a list of books currently for sale about the buffalo. We made no attempt to include books printed prior to 1950. We also decided not to include magazine articles in this bibliography. We are sure this list does not include every title about the buffalo, but it includes all those we could find.

This bibliography is divided into five categories. Parts I, II, and III include titles currently for sale arranged alphabetically by author. Part IV lists books that are not currently available. Part V is a list of publishers with their addresses.

We hope we have supplied you with enough information to order any title that is of interest to you. When available, we included book identification numbers which should be included when ordering books. To find a publisher's complete address, look up the code in Part V.

It is always easier to buy books at your local book store. But many of us live in rural areas with limited access to book stores. The bibliography should give a bookstore enough information to order the book for you. For those of you without a bookstore, the bibliography will give you enough information to order the book directly from the publisher.

The prices given are the most current available. It is possible that prices have increased on some titles. Postage must be included when ordering directly from the publishers. When the book is available in both paperbound and hardbound, that information is included.

We hope that the members of the National Buffalo Association will find this bibliography helpful. If you order books directly from the publisher, please mention that you received the ordering information from the N.B.A. Buffalo Books Bibliography.

PART I

BISON-AMERICAN

Allen, Joel A.
The American Bisons Living and Extinct. Reprint of an 1876 book. 1974. Arno $17.70 postpaid, hardbound

Anderson, Charles G.
In Search of the Buffalo, the story of J. Wright Mooar. L. V. Boling $5.95 postpaid, hardbound

Barsness, Larry
The Bison in Art: a graphic chronicle of the American bison. 1977. Northland $14.50 paperbound

Branch, E. Douglas
The Hunting of the Buffalo. #BB130. U. of Nebraska $4.25 paperbound

Cockrill, W. Ross
Husbandry and Health of the Domestic Buffalo. 1975 Unipub $25.00

Dary, David A.
The Buffalo Book. 1975. Swallow, limited edition $100.00, regular edition $15.00, hardbound; Avon $2.25 paperbound

Frison, George (ed.)
The Casper Site: A Hell Gap Bison Kill on the High Plains. 1974. ISBN 0-12-268550-4. Academic $16.95 hardbound

Gard, Wayne
The Great Buffalo Hunt. #BB390. U. of Nebraska $3.95 paperbound

Garretson, Martin S.
A Short History of the American Bison. Reprint of a 1934 book. LC 79-169759 ISBN 0-8369-5979-5 Arno $10.50 hardbound

Grinnel, George Bird
Last of the Buffalo. 1970. Reprint of an 1893 book. LC 78-125740 ISBN 0-405-02665-X Arno $7.00 hardbound

Haines, Francis
The Buffalo. 1975. Apollo $3.95 paperbound

Hornaday, William T.
Extermination of the American Bison. Reprint of an 1887 book. ISBN 0-8466-0259-8,SJS 259 Shorey $10.00 paperbound

McHugh, Tom
The Time of the Buffalo. Tom McHugh, $12.50 hardbound; U. of Nebraska, #BB685 $5.50 tentative price, paperbound

Marquiss, Toots assisted by Nellie Houck
Buffalo Cook Book. December 1977. N.B.A. $2.00 paperbound

Martin, Cy
The Saga of the Buffalo. 1973. L. V. Boling $10.00 hardbound

Olsen, S. J.
Post-Cranial Skeletal Characters of Bison and Bos. Reprint of a 1960 book. Kraus $5.00 paperbound

Park, Ed
The World of the Bison. Lippincott $7.95 hardbound

Roe, Frank Gilbert
The North American Buffalo. U. of Toronto $35.00 #17029 hardbound; $10.00 #61370 paperbound

Rorabacker, J. Albert
The American Buffalo in Transition. North Star $6.50 hardbound

Streeter, Floyd B.
Prairie Trails and Cow Towns. 1963. Devin-Adair $6.95 hardbound

PART II

BISON-AMERICAN, RELATED BOOKS

Cummings, Jean
Why They Call Him the Buffalo Doctor. Cummings, First edition $6.61 postpaid, hardbound. NOTE: Published by Prentice Hall, now out of print, but can be purchased from the author.

Easton, Robert and Mackenzie Brown
Lord Of Beasts: The Saga of Buffalo Jones. 1970 U. of Nebraska $2.95 paperbound

Grinnel, George Bird
The Cheyenne Indians 2 volumes. #071023-2 U. of Nebraska $8.50 for the set, paperbound

Grinnel, George Bird
The Fighting Cheyenne. U. of Oklahoma $9.95 hardbound

Grinnel, George Bird
When Buffalo Ran. U. of Oklahoma $3.95 hardbound; $2.95 paperbound

Haley, James L.
The Buffalo War. 1976. Doubleday $7.95 hardbound

Inman, Henry
Buffalo Jones, Adventures of the Plains. #BB513. U. of Nebraska $2.95 paperbound.
NOTE: Not in current catalog, inquire.

Jennings, Dana Close
  Where the Buffalo Roam Again: The Story of the Roy Houck Family Ranch. North Plains Press 1969 $3.00 prepaid, paperbound Houck, limited supply on hand—no longer in print.

Lee, Wayne C.
  Scotty Philip: The Man Who Saved the Buffalo. Caxton $7.95 prepaid, hardbound

Mails, Thos. E.
  Dog Soldiers, Bear Men & Buffalo Women. Prentice-Hall. $25. Hardbound. A study of the societies and cults of the Plains Indians. Descriptions of Indian ceremonies and regalia. Buffalo receive frequent and prominent mention. 1973, ISBN 013-217216-X. 013-217216-X.

Palliser, John
  Solitary Rambles and Adventures of a Hunter in the Prairies. Reprint of an 1853 book. Tuttle $5.00 hardbound

Sale, Richard
  The White Buffalo. Bantam $1.75 paperbound

Sandoz, Mari
  The Buffalo Hunters: the story of the hide men. Hastings House $9.95 hardbound; U. of Nebraska $4.50 paperbound

Schultz, James Willard
  Blackfeet and Buffalo. U. of Oklahoma $9.95 hardbound

Tibbles, Thomas H.
  Buckskin and Blanket Days. U. of Nebraska $3.95 paperbound

Vestal, Stanley
  Queen of the Cow Towns: Dodge City. #220838-2 U. of Nebraska $3.45 paperbound

PART III
BISON-AMERICAN, JUVENILE LITERATURE

Bjorklund, Lorence
  Bison: The Great American Buffalo. 1970. LC 79-105902 ISBN 0-529-00770-3,A3444 William Collins & World $5.95

Hofsinde, Robert
  Indian and the Buffalo. 1961. Grades 4-6. ISBN 0-688-31-420-1 William Morrow $5.32

Lavin, Sigmund A. and Vincent Scuro
  Wonders of the Bison World. Wonders Series. 1975. Grade 4 up. ISBN 0-396-07146-5 Dodd, Mead $4.95

McClung, Rogert M.
  Shag. Last of the Plains Buffalo. 1960. Grades 4-6. ISBN 0-688-31800-2 William Morrow $5.49

Nesbitt, Nell S.
Bumpy Bison. Grades 3-7. Binford $3.50

Scott, Jack D.
Return of the Buffalo. 1976. Grades 6-8. LC 76-19019 ISBN 0-399-20552-7 G.P. Putnam $6.95

PART IV

OUT OF PRINT BOOKS

The following books are no longer available from the publisher, or we were not able to obtain ordering information. They are included, however, as many can be located through interlibrary loan programs. They may also be found through book dealers who handle out-of-print requests.

Andrist, Ralph K.
The Long Death. 1964

Brown, Mark and W. R. Fenton
Frontier Years. 1950

Dodge, Col. Richard
Plains of the Great West. 1959

Duncan, Bob
Buffalo Country. 1959. E.P. Dutton

Jennings, Dana Close
Buffalo History and Husbandry. NBA

Long Lance, Chief Buffalo Child
Long Lance. 1928. Cosmopolitan

Mayer, Frank and Charles B. Roth
The Buffalo Harvest. 1958. Alan Swallow Sage Publications

Meagher, Margaret Mary
The Bison of Yellowstone National Park

Veglahn, Nancy
The Buffalo King. Charles Scribner's Sons

PART V

PUBLISHER'S LIST

Academic   Academic Press, Inc.
111 Fifth Avenue
New York, New York 10003

Apollo   Apollo Editions
Harper and Row Publishers
Scranton, Pennsylvania 18512

Arno   Arno Press
3 Park Avenue
New York, New York 10016

Avon   Avon Books, a Div. of Hearst Corp.
959 Eighth Avenue
New York, New York 10019

Bantam   Bantam Books, Inc.
Dept. BOW
414 East Golf Road
Des Plains, Illinois 60016

Binford   Binford and Mort Publications
2536 S. E. 11th Avenue
Portland, Oregon 97202

Caxton   Caxton Printers LTD
312 Main
Caldwell, Idaho 83605

Cummings   Mrs. Jean Cummings
Rx Ranch Enterprises
4211 S. Brooks Rd.
Muskegon, Michigan 49444

Devin-Adair   Devin-Adair Co., Inc.
143 South Beach Avenue
Old Greenwich, Connecticut 06870

Dodd, Mead   Dodd, Mead and Co.
79 Madison Avenue
New York, New York 10016

Doubleday   Doubleday Co., Inc.
501 Franklin Avenue
Garden City, New York 11530

G.P. Putnam	G.P. Putnam's Sons 390 Murray Hill Pkwy. East Rutherford, New Jersey 07073
Hastings	Hastings House Publications, Inc. 10 E. 40th Street New York, New York 10016
Houck	Mr. L. R. (Roy) Houck Triple U Enterprises Box 995 Pierre, South Dakota 57501
Kraus	Kraus Reprint Co. Div. of Kraus-Thompson Org. Ltd. Rte 100 Millwood, New York 10546
L. V. Boling	L. V. Boling-Books 3413 Tiger Lane Corpus Christi, Texas 78415
Lippincott	J.B. Lippincott Co. East Washington Square Philadelphia, Pennsylvania 19105
NBA	National Buffalo Association Box 706 Custer, South Dakota 57730
North Star	North Star Press 1400 St. Germain Street St. Cloud, Minnesota 56701
Northland	Northland Press P.O. Box N Flagstaff, Arizona 86002
Prentice-Hall	Prentice-Hall, Inc. Englewood Cliffs, New Jersey 07632
Shorey	Shorey Publications 110 Union Street Seattle, Washington 98111
Swallow	Swallow Press, Inc. 811 W. Junior Terrace Chicago, Illinois 60613
Tom McHugh	Mr. Tom McHugh 8560 Franklin Los Angeles, California 90069

Tuttle	Charles E. Tuttle Co., Inc. P.O. Box 470 Rutland, Vermont 05701
Unipub	Unipub, Inc. 345 Park Avenue South New York, New York 10010
U. of Nebraska	University of Nebraska Press 901 North 17th Street Lincoln, Nebraska 68588
U. of Oklahoma	University of Oklahoma Press 1005 Asp Avenue Norman, Oklahoma 73019
U. of Toronto	University of Toronto Press 5210 Dufferin Street Downsview, Ontario, Canada M3H 5T8
William Collins and World	William Collins and World Publishing Co. 2080 W. 117th Street Cleveland, Ohio 44111
William Morrow	William Morrow Co., Inc. Wilmore Warehouse 6 Henderson Drive West Caldwell, New Jersey 07006

Index

Technical Commission 124
Uniform Methods & Rules
& Education 122, 125
vaccine 132, 167, 239
Wichita Mountains 285
Yellowstone 282
bruises 171, 182, 203
Buffalo Breeders Assn. 349
Buffalo Chips 350
Buffalo Doctor, The 319ff.
buffalo grass 71
buffalo jump sites 335–6
BUFFALO! Magazine 107,
186, 343, 348, 350
"Buff" the bucking buffalo 261
bulls 19, 36–7, 83ff., 121, 236
feeding 240, 246–7
hunting 207
trophy head 198
burgers 183
butchering 37, 66, 181, 182ff.,
199, 241, 247, 248, 260, 264
buying 28, 87, 164
by-products 7
bits & pieces 203
bladder 203
brains 196
cape 66, 195ff.
feet 203
head & hide 195, 197–8–
9ff.
hooves 203
horns 196
horn sheath 203
legs 196
odds & ends 203
penis 203
robe, rug 199ff., 261
scrimshaw 203
sinew 203
skull 196, 201, 261
stomach 203
tail 203
tendon 203
toenail 203, 246

tongue 196
wool 203ff.

C

calcium content of meat 308,
312
calves, weaning & feeding
76ff., 85ff., 164, 232, 269
Custer State Park 293
late calves 270
National Bison Range 296
Canadian Buffalo Assn. 178
Canadian market 178, 240
regulations 178
Canadian Wildlife Service
296ff.
cancer 319ff.
American Cancer Society
324
cure? 324
vaccine? 319, 323
CapChur System (medicated
dart) 146ff.
cape, caping 66, 195ff., 199
capital gain 10
carcass characteristics 307ff.
castrating 179
Catalina Island
corrals 49
buffalo population 340
certification 111, 113
checkbook ranching 11
chemicals 74
cherry & chokecherry leaf
poisoning 69, 243, 254
Chip Flip olympics 217ff.
cholesterol content 304
chopped buffalo 185
Circle 3 Buffalo Ranch 56ff.
classification, zoologic 15
clearinghouse (NBA) 186, 352
clostridium 119ff., 166, 172
Cody, W. F., Buffalo Bill 338
co-ed herd 83ff.
Cohn, Dr. 125

fencing 17, 33, 37ff., 253
  Custer State Park 271
  hi-tensile 47
  see-thru 169
  portable 47
  temporary 47
field slaughter 181, 247, 266
flies 139, 140, 146ff.
  & pink eye 168
  biological control 140
Flocchini, Armando 229ff.,
  258, 349
  Lena 229
flukes 144ff., 168
flushing 75, 85
Fogel, Mike 107
food 183
  burgers 312, 314, 318
  chuck 314
  cooking methods 304–5,
  313
  jerky 306ff.
  patties 185, 246, 301ff.,
  311ff., 314
  pemmican 306
  roasts 314
  wet/dry weights 307
founder & bloat 70
Fraley, C. S. 72
Francis, P. B., III 324
fraud in registry 115
freeze branding 203
Furnas, E. L. 139

G

Gaede, Bob 50, 52
growth & development 36
Garretson, M. S. 337, 339
genetics 27, 86, 90
gestation 19, 37
Glosser, Dr. 127, 129
Goodnight, Charles 325ff.,
  332, 338, 340
  Mary 326

goose wing, buffalo meat
  compared to 304, 311
grandfathering-in (registry)
  111–2
grasses 70
  buffalo 71
  prairie 72ff.

H

habitat 167
  Yellowstone 279ff.
Hall of Fame, Buffalo 343ff.
Hamilton, Mort 204
handling 35, 37ff., 169, 238ff.,
  245, 249
  Custer State Park 268
metal vs. wood facilities 287–8
Hartmann, Jim 37, 56, 61, 223
hay 74, 164, 166–7
hazards 17
healing properties 319ff.
health, buffalo 119ff.
health, human 301ff., 311,
  312, 319ff.
health offices, state, phone
  numbers 165
Hebbring, J. 121, 350, 351
Heikes, Grete 204
hide hunters, 19th century
  331
hides 180
  salting 200
  shipping 201
  skinning 195ff.
  value 201, 250
history 329ff.
hoof rot 168
hoof trimming 171
Hornaday, W. T. 284, 334,
  337–8
horn measurements 212ff.
  powder 198
  registry 216
horse serum 321

hot prods 270
Houck, Jerry 61, 208
    Jimmy 61
    Kay 93
    Lila 61, 204
    Nellie 344, 346
    ranch 61, 204, 247
    Roy 16, 61, 70, 85, 87, 119,
      212, 247, 344, 345, 349,
      350
Howard, J. R. 181
hunting 65, 195, 207ff.
    arms & ammo 61ff., 208ff.
    sources 212
    Table Mt. 213
    Utah hunt 211ff.
    UUU hunt 61ff., 208ff.
hybridization 15, 19, 28, 86–7

I

immunities 162
immunology 320ff.
import requirements, Canada
    176
inbreeding 88ff., 237
Indian usage 203ff., 331, 333-
    4-5-6
injections 146ff., 171
injuries
    cuts 167
    eye 168
inspection, state laws 179
    federal laws 179
instincts re feed 163–4
intoxication 254
iron content of meat, com-
    parative 304, 312
insurance 95ff.

J

Jackson, Warren 55
Jennings, D. C. 350
Johnson, Dr. W. A., PhD
    302ff., 317

Johnstone, Robert 286
Jones, C. J. "Buffalo" 332,
    338, 340, 346
judging 29
jump sites, buffalo 336ff.

K

Kassirer, Ben & Marie 185
Klukas, Rich 127ff.
Kuhn, R. E. 95, 225

L

land 30
Laubscher, Ramona 223
leading (vs. driving) 74, 77,
    269, 286
leather 201ff.
LeFaive, Thomas 54
legislation 350
Leland, S. E., Jr. 142
lepto 167
liability 95
Lietzau, Chuck 198
    address 203
lightning 274
linebreeding 88, 269
Little Buffalo Ranch 28, 88,
    255ff., 346
loading tricks 155, 241
lockjaw 119
London, Gary 61, 80, 140
longevity 19
Lott, Dr. D. F. 49
louse control 142
lung worm 143ff.

M

Malcolm, Jon 294
managing
    Custer State Park 290ff.
    drylot 82ff.
    on pasture 74, 232ff.
    Yellowstone 278ff.
    zero management 255ff.

O

Old West Trail Foundation 350

Oklahoma Rebellion 131ff.

On the Edge of Common Sense 132ff.

overeating disease 119

ox warbles 139

P

Pablo-Allard herd 24, 296, 332, 332, 338

packer's advice to buffalo raisers 181ff.

Palmer, H. C. "Red" 155ff.

parasites 70ff., 76, 137ff., 146ff., 168, 169, 239, 244, 255

Parker, Chief Quanah 284

Park Service, Federal 127ff.

pasture 232ff., 246, 295

pasture 70ff.

   grasses 70

   poisoning 74

   rotation of 72, 232, 295

   seeding 235

Peden, Dr. Wendell 260

pedigree 112

"Pennsylvania" bison 334

Perry, Tony 252

Peterson, Lawrence J. 346, 350

pets 16, 35, 271

Pettinga, Barbara 252

Pfitzmayer, Al 208

phenothiazine 138

Philip, James Scotty 24, 232, 259, 288, 339, 340, 343–4

   Sally 344

phone numbers of state health offices 165

photographic records 191

pine needles & abortion 69

pink eye 167–8

pishkuns 336

Plains bison 16, 24, 26

playing games 163, 255

poisons 69, 74, 243, 254

   acorns 69

   cherry 254

   cyanide 254

   sorghums 69

population of buffalo

   buffalo vs. cattle, U.S. 6, 13

   Canada 296, 298

   Catalina Island 340

   Custer State Park 295

   historic 338–9, 340

   National Bison Range 295

   world, low point 332

   Yellowstone 282

   zoo 339

prairie grasses 72

pregnancy test 236

Price, Les 266, 349

prices: buffalo vs. beef 6, 186, 241, 250

   by-products 196, 200ff., 246

   hunting 207

pricing meat 241, 250

prods, hot 270

profit 7, 9, 12, 13, 19, 117, 186

promotion & publicity 217ff., 350

Propst, A. D. 49

protective feeding instincts 164

protein content, comparative 304, 312

   direct 305

putting out 12

Pyle, Dr. Robert 125

Q

quarantine 166

R

rabbit leg 266
range, eastern 334
  modern 24
  original 23
rations
  finishing 78, 80
  supplemental winter 74,
    166, 234, 245
  weaning 232
Raynolds, Dave 195, 212, 337
record animals 212ff.
  Boone & Crockett 212ff.
  NBA 214ff.
record-keeping 190ff.
  photo records 191
  registration 216
registration 90ff., 266, 268,
  352
  ABA 113
  fraud 115
  NABR 107ff.
  program 114
  records 216
  standards 111
Rehborg, Billie Ann & Raymond 344
Reynolds, Dave 91
Reynolds, H. W. 296
Rivers, Mary 343
robe, rug 196ff.
Robillard, Wayne 196
Rompun tranquilizer 152ff.
Roosevelt Conservation
  Award 350-1
rotate pastures 71, 232, 295
roundup 61ff., 85, 237
  Custer State Park 268ff.,
    290
rut (CSP) 293-4
R_x brand, ranch 320ff.

S

salt 75
  + wormer 138
  salting hides 201
Santa Catalina 24
scours 70
screw worms 139
Seabold, Paul 151
selection 88, 91ff.
senses 255, 274
serologic tests 109ff.
Seven-Way inoculum 119
sheep, similarities to 162
Sherwin ranch 24, 340
shelter 30, 33, 162
Shepherd, J. A. 93, 154
Sherwood, J. H. 286
shooting 65, 201
shots 146ff., 171
single sire mating 107, 108,
  112, 113, 114
skinning 69, 199ff.
skull 196, 201ff.
slaughter 182ff., 199, 241,
  247-8
  field 181, 247, 266
  approved slaughterhouses,
    list of 180-1
Smith, Ray 56ff.
  Robert 60
  Warren 60
Smithsonian Institution 337
sorghum poisoning 69
Spearhead (state of WY) herds
  286ff.
spirit of the buffalo 162
state health offices, phone
  numbers 165
state laws 166
stock share partnership 11
Stockstil electronic tranquilizer 157ff.
Stormont, Dr. C. J. 109, 296

Y

Z